Lumberjack Lingo

By
L. G. Sorden
And
Jacque Vallier

Tim Benge
Happy Birthday
Love Mom + Dad

12-7-92

Lumberjack Lingo

PREFACE

The lumberjack had a language all his own.

This logging dictionary is an attempt to preserve the terminology of the lumberjack in the early days of logging in the lake states.

The meaning of a word may differ greatly from state to state and even from one area or community to another.

The lumberjack was not a talkative fellow while he was working; hence, sometimes one word or phrase would take the place of many words in our present-day language. Talking interfered with his work. Even at mealtime, not a word was spoken unless it helped speed up the eating process.

The vocabulary that came to life in the logging camp consisted mainly of words that, first, were colorful; second, could be said without exercising the tongue too violently; and third, had a clear enough ring to carry easily.

Their language was a man's language, as they worked only with men. It was a rough language. It would seem that the lumberjacks were the ones who made the four-letter words popular. For the sake of propriety, offensive terms and words have been omitted.

The lumberjacks were men of many nationalities, with Swedes, Norwegians, Finns, French Canadians, and men from Maine predominating, and workers from the central and southern European countries in lesser numbers.

While these men of many nations were in logging camps doing long hours of hard work, they became a group with characteristics differing from any other group in America.

L. G. Sorden
Madison, Wisconsin

A WORD ABOUT WORDS

"Why are you collecting all that stuff?" an old lady once asked me, years ago, when I was on a word hunt. "That's all past and gone."

I do not think she wanted it all to be forgotten—she was only trying to make herself accept forgetfulness.

And the answer to her question, Why collect words? Because they are among the most distinctive and revealing expressions of any community of human beings. The kind of life they led, their work, their thoughts, the things that make them laugh or fight—all are preserved in the words they use. Other men, in other times and places, with sympathy and imagination, can bring a people back through words, making them live again.

The great era of logging is past but need not be forgotten. Logging was a man's job and the words are men's words—practical, straightforward, often rough, the humor coarse but penetrating and true. Daily life furnished many a shrewd or salty figure of speech.

A dictionary of such terms preserves better than anything else the actualities of logging and of the men who engaged in it. The reader of this book will find more than three thousand four hundred of the expressions that flourished in the Great Lakes camps for some seventy years—many that are not in our general dictionaries. Professor Sorden, and those who have helped him in the considerable labor of collecting and presenting this list, deserve both congratulations and thanks from all loggers, Great Lakers, word-lovers, and other honest men.

F. G. Cassidy
Professor of English
University of Wisconsin

In the Beginning

According to the National Forest Products Association, the first official use of the word "lumber", referring to sawn timber, appeared in the United States in Boston records of 1663.

The English used the term "lumbered up", in referring to logs, timber, and boards which littered the harbor. Later the colonists began to refer to such timber simply as lumber.

ABUTMENT A rock or timber structure built along bank of stream to keep logs in channel.

ACID WOOD Wood suitable for the manufacture of wood alcohol and other distillation products.

ADAM'S FRUIT Dried apples, pregnant women.

ADDLE To earn by labor.

A DOLLAR A THROW The cost of a whore's companionship.

ADZE A tool used in hewing logs to flatten them on one side. It had a handle long enough to enable the worker to use it while standing. Same as shin cutter.

A-FRAME GIN POLE A gin pole with a vertical portion built as an A-Frame.

A-FRAME JAMMER   Later day term for gin pole or horse jammer.

AFTER THE TURN OF THE NIGHT   Any time after midnight.

AGROPELTER   A horrible animal that lived in hollow trees mostly in Minnesota. Any lumberjack walking near its abode was killed by a falling limb.

AIR SAW   A compressed-air driven portable saw used in later-day logging operations in the lake states.

ALDER GRAB   The stem of an alder, or other small tree, which is bent over and plugged into a hole bored in a boom stick, or secured in some other way, to hold a boom of logs in shore.

ALIBI DAY   Payday in camp when many loggers developed toothaches or other ills requiring trips to town.

ALLEY or ALLEY WAY   The roofed-over space between the kitchen and sleeping quarters where meat and wood were stored. Same as dingle, midway.

ALL HANDS AND THE COOK   The call or cry that a log jam had occurred and that every one of the river crew was needed to help break it. Also used in other emergencies when help was needed.

ALLIGATOR   1. A boat used in handling floating logs. It could be moved overland from one body of water to another by its own power, usually applied through drum and cable. The propellor was encased to enable it to climb over floating logs. 2. A device, often made from the fork of a tree, on which the front end of a log is placed to facilitate skidding on swampy ground. 3. Name given to the headworks of a log-towing mechanism mounted on a heavy, large raft. The apparatus consisted of a windlass (capstan or spool) turned by horses with ropes, with a large anchor for mooring.

ALL OUT   A term used by the camp foreman when he opened the bunkhouse door and called the men to work in the morning.

A LONG ROUTE   To stay at work in one camp for a long time, perhaps several years.

—A—

AMEN CORNER a corner of the bunkhouse where the old-timers reminisced, or a place where sky pilots spoke, or entertainers told their stories.

AMERICAN LOADER—MODEL C A steam-powered log loader operating from a railroad log car.

AMERICAN LOADER—MODELS D AND E Similar to the McGiffert steam log loader.

ANCHOR BRAND STRAWBERRIES Prunes. A term used on Great Lakes lumber ships.

ANCHOR ICE Ice in shady places and long banks of streams that stayed after ice in center of the stream has broken up. An unexplained phenomenon whereby ice forms on the bottom of rapids in a fast flowing stream.

ANCHOR LINE A line attached to a small buoy and to one fluke of an anchor used in towing a raft of logs. The fluke is that part of the anchor which fastens in the ground. The anchor line is employed to free the anchor when caught onto rocks or snags.

ANGLE BAR A steel plate with a flange base, having from four to six holes, through which bolts may be inserted. Two angle bars are used to hold steel rails together at joints, one angle bar being placed against each side of the web and both bolted to it.

ANIMAL Ox, horse, or mule used for skidding and other woods work.

APRON 1. A platform projecting downstream from the sluiceway of a dam to launch into the stream logs that pass through the sluiceway. 2. A platform built of timbers at the foot of a slide that guides logs leaving the slide in the desired direction.

ARBUCKLE Coffee.

ARK 1. Where the camp stores are kept. 2. The boat or raft that followed the logs down the river on a drive. Same as wanigan, shanty boat, van.

ARSNEAU A derrick for loading logs. Same as jammer.

AS EASY AS FALLING OFF A LOG   A term that came directly from the woods and the log drives.

ASSORT UP   Inventory in a lumber yard.

ATE BIG   Good grub and lots of it.

A-TENT   A very small tent for one or two men consisting of canvas in an A-shape and was used by the early land lookers and timber cruisers.

AXE or AX   The conventional chopping tool of the lumberjack. He took great pride in his axe and did not leave it sticking in a tree or stump overnight, for he believed that to do so would bring him bad luck. There are three kinds of axes: a single-bitted or poll axe, a double-bitted axe, and a broad axe. In very early logging, saws were not used, only axes. Even in the state of Maine and other New England states, saws were not generally used for cutting down trees until about 1890.

AXE HANDLE PUNCH   A flat-edged punch used to drive broken axe handles out of the eye of the axe head.

AXE HANDLE TYPES   1. Straight Handle 2. Curved Handle 3. Fawn (deer) Foot 4. Scroll Knob 5. Swell Knob

AXE HELVE   An axe handle.

AXE MAN   A woods worker whose main job was using the double-bitted axe. Also the member of a survey crew who blazed a line for the surveyor. A faller.

AUGERMEN   Workers who drilled holes with hand augers into which were inserted the raft pins in the construction of timber (log) and lumber rafts.

Loggers' Table

At 5 A.M. the gabriel was blown or an iron was struck with a stamping hammer and you crossed the dingle or raced to the cook shack, if the buildings were separated. You took the same place at the same oilcloth-covered table for all meals. You had to hurry or be the last one to leave the table. Some men could eat and run in five minutes; everyone was through in fifteen minutes. Chewing your food was a mistake—you had to learn to bolt it. The cookee had the table set with tin plates and cups, knives, forks, and spoons, and no one left the table hungry.

Pies were made for the midday meal and sauce was served for supper. One camp cook told us his routine: custard pie on Monday, raisin pie on Wednesday, dried apples with raisin sauce for Thursday supper, and prune pie on Saturday. He also said prunes made better apple pies than peaches did.

BABE   The name of mythical Paul Bunyan's Blue Ox.

BACK CANT   Reverse use of cant hook to hold log back instead of the usual use to roll it forward.

BACK CUT   In sawing down timber, the cut put into the tree after the undercut.

BACKFIRE   A fire set in front of the oncoming fire to stop the main fire.

BACK FORTY   Way back in the brush—a long haul.

BACK HAUL   The line that returns the chokers to the woods after a turn of logs has been pulled in.

BACK HOUSE   Outdoor, log privy.

BACK PUMP   A portable water tank of later years used to extinguish forest fires.

BACK SPIKER   One of the members of a crew who completes the spiking of rails to cross ties after the track has been laid by the steel gang.

BACK TIMBER   Timber located inland from waterways and which required sleigh hauls for shipment downriver. (The converse of shore timber.)

BADGER   Brand of smoking or chewing tobacco used by the lumberjacks. Same as Wilder.

BAG BOOM   An open boom used to impound logs at the mouth of a stream.

BAGNIO   A girl house. Sometimes called a boarding house. A famous one in the upper peninsula was known as the Klondike, and the ladies were called Klondikes.

BAGS   Same as bag boom.

BAKE A BATCH OF ROLLS   Prostitute jargon for sexual intercourse.

BAKE OVEN   A cruiser's tent with one sloping side and two ends.

BAKE-OVEN TENT   A tent used by the early land lookers and timber cruisers having one sloping-side and two ends, like an old-fashioned tin bake oven used before an open fireplace. A log fire toward the open side made for great warmth.

BAKER   A large stove with a series of ovens placed one above the other; used for baking.

B & L BLACK   A popular chewing tobacco.

6

−B−

BALD-FACED WHISKEY A potent, cheap grade of whiskey.

BALDWIN CONSOLIDATION A locomotive used in lake states logging operations.

BALDWIN TANK ENGINE A small locomotive made by the Baldwin Locomotive Company having a water tank located over the drivers for traction.

BALLAST CARS Railroad cars similar to gondola cars used to haul ballast for railroad track fill operations.

BALL HAMMER A small iron hammer with a flat head on one end and a pick end on the other. Carried on the harness hames, it was used for knocking out balls of ice and snow that had formed under the shoes of horses while they were working. Oxen did not have this trouble because of their cloven hoofs. Same as peen hammer, snow knocker.

BALL HOOTER One who rolls logs down a hillside.

BALLOON A pack; a bedroll.

BALL THE JACK To travel fast.

BALSAM 1. Short for balsam fir. 2. The gum or pitch from a fir.

BALSAM MYRRH 1. Pitch from the balsam tree used for healing wounds. The way to treat a split hand was to run hot balsam pitch into the wound and light a match to it. The wound healed and didn't even leave a scar. 2. A liquid patent medicine, a type of liniment.

BAND SAW A saw in the form of an endless belt running over pulleys. Used in all large sawmills.

BAND SAWYER Operator of a band saw in a sawmill.

BANGOR SNUBBER Same as Barienger Brake.

BANICK From Scotch word bannock meaning biscuit, a hand bread. A type of bread used on a log drive made with flour, baking powder, and water. It was not very appetizing.

BANJO A number two shovel.

BANK   To pile up logs on a landing for transportation to the mill.

BANK BEAVERS   Rivermen or river rats driving the rear on a log drive.

BANKED LOGS   Logs piled at the edge of a stream or lake ready to float on rising spring waters.

BANKERS   Lumberjacks who pile logs on the banking ground.

BANKING   Decking or piling logs on the river bank. Logs were later dumped into the water and floated downstream to the mills.

BANKING GROUND   A landing or rollway where logs are piled.

BANK SCALE   Number of board feet of logs piled at the banking ground or landing.

BAPTIST CONE   Used in skidding operations, a steel cone placed over the front end of a log to prevent it from digging into the ground. Invented by William Baptist of Green Bay, Wisconsin.

BARBER CHAIR   A stump with part of the tree left due to the tree's splitting when falling. Same as tombstone.

BARBER POLE   A sawed tree fallen on another, bending it over.

BAREFOOT   How a lumberjack felt when he wore smooth-soled shoes instead of calked boots.

BAR HORSESHOE   A horseshoe with a metal plate, put on where a horse had a sore heel or a split hoof. A pad could be put between the sore and the shoe to protect the sore.

BARIENGER BRAKE   A series of drums wound with heavy cable and operated by a lever. It would hold or break the heaviest sleighload of logs on downhill grades.

BARK BLAZER   A sharp axe or hatchet used to cleave a portion of bark from a tree.

BARK DRAY   A dray used to haul hemlock bark out of the woods. Bark was cut in four-foot lengths. Hemlock bark was used in the tanning industry.

8

BARK EATER   1. A lumberjack. 2. A sawmill hand.

BARKED   Same as sniped.

BARKER   1. One who peels bark in gathering tanbark. 2. The tool to peel bark. Same as peeler, spudder.

BARK HACKS   Bark marks made to identify ownership of logs in a river boom. Same as bark marks or side marks.

BARK HACKS   Bark marks made to identify ownership of logs in a river boom. Same as bark marks or side marks.

BARKING   Same as sniping a log.

BARKING AXE   A small, short-handled, broad axe for ringing bark before peeling with a spud or barking iron.

BARKING IRON   A tool for peeling bark off logs. Same as bark spud, spud.

BARKING SPUD   To remove bark from logs. Same as spud or barking iron.

BARK IS SLIPPING   A good time to peel poles or pulp sticks.

BARK LADDER   A platform or rack mounted on wagon or sled and used for hauling tanbark.

BARK MARK   A mark stamped into the side of a log to denote ownership. Same as side mark, water mark, log mark.

BARK MARKER   One who cuts the bark mark or water mark on logs.

BARK PEELER   A cant hook that did not function properly in rolling logs. It would crease the back of the log instead of taking a firm hold, because the point of the hook was dull.

BARK SPUD   A tool for peeling bark off logs. Hemlock bark was sold to tanneries. Same as barking iron, spud.

BARKY STRIPS   Crating material.

BARN BOSS   A man whose duty was to care for the horses and stables in a logging camp. Same as feeder.

BARN BOY   A barn boss's helper.

BARNDOOR GATE   In a logging dam sluiceway, a swinging door attached by hinges to the side of the sluice so that it can be swung across the opening to prevent the outflow of water.

BARNHART LOADER   A steam-powered log loader operated from a railroad log car.

BARN LANTERNS   Same as Teamster's lanterns.

BARREL HOLES   Round holes in the shanty roof into which barrels were set upside down with holes drilled in the side for ventilation. It replaced the smoke hole in the camboose.

BAR ROOM   That part of a camp where the loggers slept. Term used occasionally in lake states, more often in New England.

BAR ROOM MAN   A chore boy, bull cook, flunky.

BASE CAMP   Same as headquarters camp.

BASE LINE   The beginning line in a wood survey or timber cruise.

BASKET RACK   A sled rack with high staked sides on which pulpwood is piled.

BATEAU   Type of boat used on river drives. It was a French boat used first in fur trading, built to stand heavy work, especially hard to tip over. Flat-bottomed, tapering toward the ends, it was sometimes forty feet long.

BEADS   Chain used in loading logs.

BEAM   The heavy timber holding the runners of a sleigh in place near the center of the runner.

BEAM DRIFTED   A cruiser's expression for a line slightly off the correct one in order that certain trees would be included in the final tally for cutting.

BEAM SCALE   A weight scale having a metal balance bar. (Used to weigh all bulk food arriving at camps.)

BEAN HOLE A hole in the ground in which beans were baked, used as an oven in early logging camps. Hardwood coals were raked into it and on top of the bean kettle, which had a tight lid; then it was covered with dirt. The beans were said to have tasted extra good.

BEAN HOLE BEAN POT Same as bean pot.

BEAN HOUSE The foreman's office in a logging camp.

BEAN POT Cylindrical cast-iron pot with a heavy, tight-fitting lid in which beans were baked in a bean hole.

BEAN SHEET A workman's daily record sheet.

BEAN WAGON Supply wagon or tote wagon.

BEAR BAIT Spoiled meat fed to lumberjacks so they would not eat too much.

BEAR PAWS A type of snowshoe used at early logging camps.

BEARING TREE A tree blazed and marked by government surveyors to indicate the location of township or section or quarter-section corners. Same as witness tree.

BEAR TRAP 1. A tricky situation. 2. An early gate mechanism used in dams to back up water when driving logs down a river. 3. A tangle of logs that may roll and kill or injure the bucker.

BEAT CAMPS Small camps located every seven to ten miles along rivers to accommodate drivers and other river hogs. It held up to thirty men.

BEATS During a drive the river was divided into beats or sections, and a small crew would handle that portion of the drive.

BEAVER A poor chopper. Same as wood pecker.

BEAVER MEADOW A grassy meadow produced along driving stream as a result of beaver dams' impounding water and killing off the trees. In such places grasses would grow abundantly and were used for feed and bedding for work animals in camp.

BED DOWN   1. To move into a logging camp with the intention of staying. 2. To put hay or other materials on the floor of the stalls in a horse barn on which the animals sleep.

BEDRAME TIMBERS   Large, heavy timbers used as base frames in sawmill construction.

BEDROLL   Blankets and perhaps other gear carried by a logger from one camp to another. Not often used, as bedding was furnished by the logging camp.

BEECHNUT   A popular smoking tobacco.

BEEF SLOUGH   Beef Slough was a protected harbor near the confluence of the Chippewa and the Mississippi Rivers. This slough offered the only practical place for Mississippi millmen to culminate their Chippewa drives and to sort and store their logs in preparation for rafting to their mills.

BEEF SLOUGH WAR   The fight between the lumbermen of the Chippewa River Valley and the "downstream" lumbermen along the Mississippi River from the confluence of the Chippewa River and the Mississippi River as far south as Rock Island, Illinois. The so-called "war" was over the right to use Beef Slough as a booming grounds.

BEGGES or BAGGIES   Rutabagas.

BELL BUTT   A log with a sharp taper on the end like a bell.

BELL OX   Foreman. A nickname given to the camp foreman because he was to the crew what the leader or bell ox was to the herd when it was turned out to pasture.

BELLY   1. Slack in a cable or towing line. 2. The middle of anything.

BELLY BURGLAR   A poor cook. Same as belly robber.

BELLY ROBBER   A name often given to the cook, especially if he was a poor one.

BELLY WRAPPER   A chain around the center of a load of logs to hold it firmly to the sleigh.

BELOW   Downriver.

BENCH HATCHET Same as shipwright axe.

BENCH MARK 1. A survey reference point, sometimes used to mean the starting point. 2. A permanent object indicating elevation and serving as reference in topographical surveys.

BENDS & RAPIDS CREW Those men in a driving crew who kept the logs going downstream.

B.F. Standard abbreviation for board foot.

BIBLE POUNDER A street corner preacher; reformer. These men, talking without a pulpit, often emphasized their talks by pounding the Bible they carried.

BICYCLE A traveling two-wheeled block used on a skyline cable in skidding logs; generally used by steam power.

BIG AS A BEAR AND LAZY AS TWO A very lazy lumberjack.

BIG BERTHAS butt logs.

BIG BLUE A log larger at one end than at the other. Same as blue, blue butt.

BIG BULL A camp foreman.

BIG BURN An extensive forest fire.

BIG FORTY When a lumberman bought a forty acre tract of timber and then proceeded to log off the whole township, it is said that he had purchased a "Big Forty".

BIG PUSH The camp foreman. Same as push.

BIG ROLL The largest gate in a logging dam. Large enough for the river wanigan to pass through.

BIG SHOTS Managers from the main office.

BIG STICKS The woods.

BIG WHEELS A pair of wheels, usually ten to fourteen feet high, used for transporting logs. Same as katydid, logging wheels, sulky, timber wheels, high wheels.

BIGWIG   A foreman; boss.

BIG WOODS   Vast forests of the lake states which were supposed to never "run out" of white pine.

BILL RAFT   Raft timbers cut to certain sizes by order.

BILTMORE   A calibrated rule stick used to estimate height and diameter of standing trees.

BIND   1. To get into a hassle or argument. 2. To get a saw stuck in sawing logs or trees. 3. Short for corner bind.

BINDER   1. A springy pole used for tightening a chain binding together a load of logs. 2. A cross pole on a log raft held in place by binding pins.

BINDING CHAIN   A chain used to bind together a load of logs. Same as wrapper chain.

BINDING LOGS   Logs placed on the top of the chain binding a load in order to take up the slack.

BINDING ON   Securing a load of logs on a sleigh with a belly wrapper chain.

BINDING PINS   Wooden pins used to wedge a binder in place in a log raft.

BINDING POLE   A pole device used to bind a load of logs to the sleigh.

BINDLE   A blanket roll. A lumberjack sometimes brought his belongings wrapped in his own blanket when he came to camp. Used in the East and in early logging in the Middle West.

BINDLE STIFF   A jack carrying his bindle. See bindle.

BIRCH BARK SNUFF BOX   Small, covered birch bark boxes whittled by jacks to hold their daily supply of snuff.

BIRCH HOOK   Same as pulp hook. Bolt hook.

14

BIRD CAGE   Various-shaped screen covers fitting over the smokestack of logging locomotives to prevent sparks from causing forest fires. A spark arrester.

BIRDSEYE TENDERLOIN   Ox meat showing marks of bull whacker's goad.

BIRL   To rotate a floating log by treading upon it to find the water mark.

BIRLER   A river pig who treads on a log in the water to rotate it and thus find the end mark or the side mark of ownership.

BIRLING   The logger's game of rolling often played on log drives. Two players wearing calked shoes spun a log to see how long their opponent could stay on without being thrown into the water.

BITCH CHAIN   1. A chain attached to a trip hook, used on eveners or whiffle trees when horses hauled logs out of the woods. 2. A short chain with hook and ring, used for fastening the lower end of a gin pole to a sled or car when loading logs. 3. Any chain used in loading logs.

BITCH LAMP   A large lantern-type lamp. Lumberjacks used to say, "If you had to get out of bed on a cold winter night, light this lamp, harness a team, hand pump a big wooden water tank on a sleigh filled with ice water, and then go slowly over a logging road, dropping the water on the sleigh runner tracks until the big tank was empty, I think you would call it a 'bitch lamp' too."

BITCH LINE   Same as bitch chain.

BITCH LINK   A pear-shaped link on the end of a chain. It was larger and heavier than the links of the chain.

BLACKBIRD   A man proficient in riding logs on a river drive.

BLACK CROW   A stove black (polish) used in the early camp days to preserve the outer finish of the iron stoves.

BLACKJACK   Coffee.

BLACK KNOT   Fungus infection at spot of old knot where the tree self-pruned.

BLACKSMITH TONGS   Various-shaped, pincer-like tools used by a blacksmith when handling hot metal at the forge or anvil. Different types were: hammer tongs, hoop tongs, round bits, square bits, flat bits, crooked bits.

BLACKSTRAP or BLACKSTRAP SYRUP   Molasses used on the morning pancakes.

BLACK WATER   A disease of horses accustomed to work and exercise but left tied up in the barn for a couple of days. It made their legs swell. Same as Monday morning leg.

BLANKET FEVER   Referring to condition of lumberjack who stays in bed in the morning after the other men are up. Generally a lazy lumberjack.

BLANKET HOIST   A punishment as well as a game. If a jack got out of place, other jacks would put him in a blanket, get hold of the sides of the blanket, and hoist him in the air a few times.

BLANKETS   Pancakes.

BLASTING WEDGE   A heavy, round wedge approximately a foot long designed to hold a charge of black powder. Inserted into logs too large to skid to blast the log in two lengthwise for skidding. A red rag was affixed to the wedge so that it could be retrieved from the snow.

BLAZE   1. To mark a tree by chipping off a piece of bark or wood to indicate a boundary, trail, road, or description or land. 2. A small forest fire.

BLAZE MARK   The spot made on a tree by blazing, to indicate a boundary, trail, road, or description of land.

BLINDERS   Flaps attached to the bridle to prevent horses from seeing objects at the side.

BLIND PUNK   A black streak in the wood, looking something like a knot.

—B—

BLOB   To make a bad woods mistake, as: "He made a Blob".

BLOCK   1. A boom filled with logs ready to be towed; a section of a log raft, six of which make an average tow. The block was developed by W. J. Young of Wisconsin in 1875. 2. A pulley (to the greenhorn). 3. Also many pulleys into which rope or cable is run.

BLOCK UP   To acquire stumpage or to square off large tracts of timberland.

BLOW   A dam is said to blow when it gives way and the head of water is lost.

BLOW DOWN   Trees that have been blown down by the wind.

BLOW HER IN   Spend your stake in the city, usually on one big spree after a winter's work in the woods.

BLOW UP   A forest fire which blazes up very suddenly and rages out of control.

BLOWIN HER IN   Cutting a certain forest stand.

BLOW MY STACK   Spent my money! Or, Got mad!

BLOW MY STAKE   Spend my winter wages on one big "fling" in town.

BLOWING FRESH   A stiff wind.

BLUE   A twelve foot log larger at one end than at the other. Prone to roll faster on one end. Same as blue butt, big blue.

BLUE BUTT   Same as big blue, blue.

BLUE BUTTED SCHOOL MARM   A crotched log with a large, swelled-up butt end.

BLUE DEVIL   A rutter used in cutting ruts for the ice road. They were generally painted blue.

BLUE GOOSE   A type of rutter used to cut ruts in the ice so sleigh runners could follow.

BLUED LUMBER   Lumber stained blue by fungi.

BLUE JACKETS   Body lice. Same as crumbs.

BLUE JAY   A man who keeps the sleigh road in good condition. Same as road monkey, hay man on the hill.

BLUE NOSES   Lumberjacks from Nova Scotia.

BLUE VITRIOL   Copper sulfate to kill body lice.

BLUING   The result of fungus attack, which turns the sapwood of certain trees blue.

BOARD FOOT   A measure of saw logs and saw timber. Actually it is a board one inch thick, twelve inches wide, and twelve inches long.

BOARDING CARS   Railroad cars converted into sleeping and eating cars for a portable logging camp operation.

BOARDING HOUSE MAN   A cook.

BOARDING HOUSE MYSTERY   Hash.

BOARD WITH AUNT POLLY   To draw insurance for sickness or accident

BOAT CREW   Six men employed to man a bateau.

BOAT PACK   Cook's gear in a river wanigan.

BOB   A two-runner sled for hauling logs out of the woods. A drag sled, dray, lizard, skidding sled, yarding sled.

BOBBER   A water-soaked log lying on the bottom of a river or lake, or a partly sunken log. Same as dead head or sinker.

BOB LOGS   To transport logs on a bob or dray.

BOBSLED   A sled used for hauling logs. Same as logging sled. A true lumberjack, however, used the word sleigh rather than bobsled or logging sled.

BOBTAILED CREW   A crew of misfit loggers usually found in a haywire camp.

—B—

BODY LICE Human parasites common in many camps. Also called crumbs, blue jackets, and gray backs.

BODY WOOD Cordwood cut from whole trees or limbs as distinguished from slab wood.

BOGS Individual rafts of 16,000 to 18,000 pieces of lumber.

BOG SHOE Same as swamp shoe.

BOGUE AWARD Named after George M. Bogue who worked out rate differential tables to govern lumber shipments.

BOHUNK Any person from southeastern Europe, generally Austria or Bohemia.

BOILED COLLAR A white dress shirt.

BOILER A bum cook. One who doesn't make tasty or well-seasoned food. Generally boiled his food. See sizzler.

BOILER HOUSE Building at a sawmill where the boilers are located and the steam to run the mill is generated.

BOILING UP, BOILING OUT, or BOIL UP Washing one's clothes; sometimes to delouse one's clothes by boiling them. Early camps were often infested with body lice and bedbugs. The lumberjacks tried to kill them by boiling their clothes when the clothes were washed. Boil up day was always Sunday.

BOIL UP DAY Same as boil up.

BOLE Base end of a tree.

BOLSTER The heavy timber upon which the logs rest on a logging sleigh. Same as bunk.

BOLT A short segment of tree stem from two to five feet long, sometimes split. Used as primary raw material by wood turneries, shingle and stave mills, and other specialized wood-using plants; or used as mine props. Same as shingle bolt, spoke bolt, stave bolt.

BOLT HEADER 1. Metal plate with various-sized and various-shaped holes for forging bolt heads; sometimes called a bore. 2. A blacksmith's hand-made tool for putting the heads on bolts.

BOLT HOOK A steel hook used to handle pulpwood. Pulp hook, birch hook.

BOOM 1. Logs chained together at ends to form a corral to hold logs in water until ready for refloating, reshipping, or sawing. 2. A group of logs pulled together by heavy boom sticks for transporting by water.

BOOMAGE 1. Toll for use of boom. 2. The area where the boom is tied.

BOOM AUGER A large auger (about three inches in diameter) used to bore holes in boom sticks through which chains were passed. Chains held boom sticks togther to form a boom.

BOOM BOOKS Office record books of the boom works.

BOOM BOSS Boss or man in charge of the entire boom works.

BOOM BUOY A heavy weight used to anchor booms in deep water. Same as boom stay.

BOOM CHAIN A short, heavy chain with a steel bar on the end which draws the chain through a hole in the log used in fastening boom sticks together, end to end, in a boom. Bar holds the chain from slipping back.

BOOM COMPANY A corporation engaged in handling floating logs, and owning booms and booming privileges.

BOOM CREW A crew of men whose work was to form, maintain and move log booms on water.

BOOM DOGS Iron wedges about 5-½ inches long connected by 4–8 or more links of chain. Wedges were driven into log ends chaining them together to make a floating log fence or boom.

—B—

BOOMER A man who doesn't stay long on any job, generally three days to three weeks. One who works just long enough to get a liquor stake and the "wrinkles out of his belly". Same as drifter.

BOOMING GROUND An enclosure near the mouth of a river where logs are sorted as to ownership or held for future disposal. Same as boom works.

BOOM JUMPER A motor boat designed to ride over loose floating pulp and logs without fouling propeller or rudder.

BOOM LAKE Any lake where logs are stored in rafts or booms.

BOOM LOG One of the logs in a boom. A boom stick.

BOOM OF LOGS The logs enclosed within a boom awaiting reshipping or manufacture.

BOOM CUT To catch wood in a boom as it enters a lake and to tow the wood to the outlet for sluicing.

BOOM PIN or **BOOM PLUG** A wooden plug used to fasten to boom sticks the chain or rope that holds them together.

BOOM RAT One who works on a boom.

BOOM STAY A heavy weight used to anchor booms in deep water. Its position is indicated by a pole or float attached to it. Same as boom buoy.

BOOM STICK One of the logs in a boom. A log that is chained to other logs to form a boom.

BOOM TORCH A torch light used in night operations at a sorting gap.

BOOM WALKER A watchman whose job it was to walk along a log boom taking care of one or more fin booms by clearing away logs or drift that would catch on fins or braces. It was his job to keep booms in general good repair.

BOOM WORKS An enclosure near the mouth of a river where logs are sorted as to ownership or held for future disposal. Same as booming ground, sorting works.

BOOT GREASE Jack's water-proofed their leather boots by applying a mixture of hot tallow, beeswax and a little lampblack. The mixture was called "Lickdob".

BOOK JACK A device made of a crotched limb or board to pull off leather boots. Also a handy device in roughhouse melees.

BOOZER A heavy drinker. Whiskey was not allowed in logging camps.

BOSS 1. Normally the man in charge of a logging operation. 2. The name for the starting bar or push. This bar broke the sleigh runners loose from the ice after they had sat for a period of time.

BOSS OF THE ROB SHOP A clerk.

BOTTES SAUVAGES Homemade shoe pacs of the French Canadian loggers.

BOTTLE BUTTED A tree greatly enlarged at the base. Same as churn butted, swell butted.

BOTTLE HOOK A wire hook attached to a bottle, usually a quart whiskey bottle, in which was kept kerosene used on the saw to prevent pitch from gumming up a saw. It was hooked onto a tree branch handy for the sawyer.

BOTTOM Soggy, lower levels in swamps that had to be frozen before any winter work could be done in swamp areas.

BOTTOM LOADER A member of the loading crew who attached the tongs onto the log for loading. Same as ground loader, hooker, hooker on, sender, send up man.

BOTTOM SILL The bottom log on a dam. Same as mud sill.

BOW BOAT A small boat pulled at the bow end of a lumber or log raft used as a steering aid for the raft.

BOW MAN The log driver who sat in the forward end of the bateau.

BOW SAW Saw with bow-type handle, generally used to cut pulpwood. Same as Finn saw, Swede saw.

—B—

BOX END OF A PIT SAW End of the saw held by the bottom sawyer in a pit saw operation.

BOX UP THE DOUGH To cook.

BOX STOVE A large, rectangular, cast iron stove used to heat camp buildings.

BRACES Suspenders to hold up the lumberjack's stag pants. Galluses.

BRAG LOAD An extremely large load or record load of logs on a sleigh. Often loaded especially for a photograph.

BRAIL A boom filled with logs ready to be towed; a section of a log raft, six of which make an average tow. The brail was developed by W. J. Young in 1875.

BRAILING CREW A crew of 16–18 men who put logs onto rafts at a sorting boom.

BRAINS A man from the head office, usually the company president or some head office official.

BRAKE SLED A logging sleigh so constructed that when the pole team held it back, a heavy iron on the side of each runner of the forward sled was forced into the roadbed and held the sleigh.

BRAKIE A brakeman on a logging railroad.

BRANCH ROAD A road leading off the main road.

BRAND The mark on a log to identify its ownership. Same as log mark.

BRANDING AXE or BRANDING HAMMER A tool for marking ownership of logs. Same as stamping hammer.

BRASS CHECK A capitalist newspaper; any publication not friendly to the I.W.W. (International Workers of the World).

BRASS COLLAR The man who owned the trees being cut, or the man for whom the logs were being cut.

BRASS HATS Directors of a company. Important stockholders or owners of the logging operation.

BRASS MONEY Tokens used instead of money.

BRAZEL A large moldboard (also spelled mouldboard) type of snow-plow.

BREAD SHOVEL A wooden, short-handled, flat shovel used to remove bread pan from ovens.

BREAK A HAM STRING To do one's best.

BREAK A JAM To start in motion logs that have jammed during a river drive.

BREAK A LANDING or **BREAK A ROLLWAY** To roll a pile of logs away from a landing on a river bank into the water.

BREAK A SKIDWAY To roll logs off a pile to be hauled away.

BREAKING A CENTER Dislodging key logs hung up on an island or other obstruction when they were being floated downstream.

BREAKING DOWN LOGS Sorting logs at a rafting gap or sorting boom.

BREAK OUT 1. to start a sled whose runners are frozen to the ground. 2. To open a logging road after heavy snowfall.

BREAKING PLOW A wooden or metal-beamed plow used to loosen earth before using a slusher to move or remove earth in road-making.

BREAK THE LOAD To start a sleigh with runners frozen to the ground.

BREAK UP Closing of a logging camp at the end of a season in spring of the year. Ice on roads melted due to warm spring season.

BREAST 1. The fore part of a jam. 2. The cutting edge of a cross-cut saw.

BREASTED SHINGLES Shingles cut from a shingle block by using a splitting froe.

24

—B—

BREAST LOG   A heavy log placed at the edge of a landing for a protection in loading operations.

BREASTWORK LOG   A log placed on the lower side of a skidding trail to hold the logs on the trail while being skidded. Same as fender skid, glancer, sheer skid.

BREECH LOADER   Bunks in a bunkhouse set parallel to the walls so that jacks had to crawl into the bunk from the side.

BRIAR   A crosscut saw.

BRIAR PIPE   A popular logging camp pipe made of briar root. By no means an exclusively lumberjack term.

BRIDGE AUGER   Same as ship auger.

BRIDLE   A special hitch for oversized logs in which chokers hooked into each other when one choker was too short to reach around log.

BRIDLE CHAIN   Chain wrapped around a sled runner to serve as a brake when going downhill on iced roads.

BRINGING DOWN THE DRIVE   Drivers or river pigs sending a mass of logs downriver.

BRINGING UP THE REAR   Same as sack the rear.

BROAD ARROW MARK   Mark used in colonial times by Queen Anne's surveyor to reserve such trees for the royal navy. These trees were said to be the finest white pine in New England and their marking was one of the causes of the Revolutionary War. Similar to an X-tree.

BROAD AXE   A large-sized axe for squaring timbers and railroad ties. The Scandinavian and French used it to a great extent in fitting logs together when putting up substantial buildings or in smoothing the interior walls. It had one straight edge, and the handle was offset to one side.

BROAD AXE BRIGADE   Men who cut ties in woods.

BROAD HATCHET   Same as shipwright axe.

BROAD LEAF   A hardwood tree, deciduous as opposed to the conifers or softwoods. Same as hardwood.

BROKE CAMP   to leave camp after finishing work for the winter.

BROKEN AXE HANDLE PUNCH   A blunt-end chisel used to drive out the remains of a broken axe handle in the "eye" of an axe.

BROLO   Any large stick or club for pounding horses to make them pull harder or to obey orders. Same as two hander.

BROOMAGE   Standard log lengths were often 16 feet, 6 inches in length. The additional 6 inches of wood was allowable by governmental regulations to compensate for damage in river driving and was referred to as "Broomage".

BROOMED LOGS   Logs battered up on their ends during a log drive.

BROW LOG   Any log used as a prop at a landing or log dump where logs are being rolled on cars or sleighs.

BROW LOGS   Logs on each side of the track at a loading landing.

BROWNHOIST LOCOMOTIVE CRANE   A self-propelled crane (derrick) or log loader equipped to run on railroad tracks.

BROWNING LOADER   A steam-powered log loader which was self-propelled on railroad tracks.

BROW SKID   A large log placed on a rollway next to, and parallel to the railroad track to prevent logs from rolling out of control onto the track.

BRUSH   Term used sarcastically by early white pine loggers to describe cedar and tamarack trees.

BRUSH A ROAD   To cover with brush the mudholes and swampy places in a logging road to make it solid.

BRUSHED   Crazy.

BRUSH HOOK   Same as brush scythe.

—B—

BRUSH OUT   1. to clear away the brush from a survey line or logging road. 2. To clear space in which to pull a saw or swing an axe safely.

BRUSH SCYTHE   A heavy-bladed, short handled (5 feet) scythe used in cutting brush from roads and skidways.

BRUSH SNOW FENCE   A snowbreak to protect a logging road; used most commonly on wide marshes. It consists of brush set upright in the ground, before it freezes, to catch the snow.

BUBBLE CUFFER   One who birls or rolls in a long birling game.

BUCK or BUCK IT UP   1. To cut a tree into proper lengths after it has been felled. 2. To yard felled trees in the woods.

BUCK BEAVER   1. Boss of crew that cut trees and brush off a new road being built. 2. Assistant to the camp boss.

BUCKER   One who saws felled trees into logs.

BUCKET BRIGADE   Tin buckets passed along lines of men to douse a sawmill or other small fire. Because fire buckets were cone shaped, they had to be hung up since they would not stand. The bottom was pointed so that they could not be used for other purposes and were therefore always available.

BUCKIN BOARD   Same as bucking board

BUCKING BOARD   A bulletin board on which a list of crews was posted. Superintendent pitted one crew against another in this manner to get a greater amount of work out of them. This practice was much despised by some workers because the men felt the competition drove them to work too hard.

BUCKING CREW   A crew of jacks cutting felled trees into logs.

BUCK SAW   Also called a wood saw. A saw set in a deep H-shaped frame, used mostly for sawing firewood.

BUCK SKIN   1. A log that has no bark on it. 2. An Indian lumberjack.

BUCKWHEAT   1. A tree felled so that it lodges against another tree. Same as hang up, lodge. 2. A grain from which buckwheat cakes are made.

BUCKWHEATER   A tenderfoot or inexperienced lumberjack.

BUCKWHEAT PINE   1. Trees that taper very rapidly and knots show from the ground. Not of much value. Sometimes occur as double pine or triple pine. 2. A young white pine with a large, low top or crown.

BUGGY   A two-wheel cart or truck used for hauling lumber at the sawmill. Same as dolly, lumber buggy.

BUGGY LOADERS   Same as trucksters.

BULL   1. The boss of a camp or a logging operation. It is a term that, when used as a prefix, except in the case of bull cook, denoted the superlative in size, power, or authority. Same as head push. 2. A tree with limbs almost to the ground.

BULL BUCKER   A man in charge of a crew of fallers and buckers.

BULL BLOCK   The main pulley-block used in high-lead logging.

BULL CHAIN   An endless chain with dogs on it used for hauling logs from a hot pond or from a load into the sawmill. Same as jack chain.

BULL COOK   The chore boy around the camp who cut fuel, filled wood boxes, swept bunkhouses, washed blankets, fed pigs; he was often the butt of camp jokes. But when he rang the gong in the morning, all the men had to get up. So called because in the early camps his first chore in the morning was to feed the oxen. Same as bar room man, chore boy, flunky.

BULL DURHAM   A popular smoking tobacco.

BULL EDGER   Main edging saw in a sawmill.

BULL 'EM THROUGH   To hurry. Same as rawhide.

BULL ENGINE   A logging locomotive.

BULL HORN   Same as dinner horn.

BULL OF THE WOODS   The camp foreman, logging superintendent, or boss of a logging crew. Same as bully, head rig, king pin, main say, top man, woods boss.

—B—

BULL PEN   The bunkhouse where lumberjacks slept. 2. The store section in a camp wanagan or wanigan.

BULL PINE   A pine that has never been cut. Generally referred to a pine standing alone.

BULL PUNCHER   A driver of oxen. Same as bull skinner.

BULL ROARER   A jack who roared when singing the bunkhouse ballads.

BULL ROPE   Decking chain.

BULLS   Often used to mean oxen used in hauling, whether male or female.

BULL SAPLING   A vigorous young tree or sapling.

BULL SKINNER   A driver of oxen. So called because the driver prodded the oxen with a sharp goad stick. Not used much after 1895. Same as bull puncher.

BULL WHACKER   An oxen teamster of the early logging days.

BULL WHIT   Boiled beef.

BULLY   A camp foreman or boss of a logging crew. Same as bull of the woods, head rig, king pin, main say, woods boss.

BUMMER   A small truck with two low wheels and a log pole, used in skidding logs. Same as drag cart.

BUMMERY   Outsider's or nonsympathizer's name for the I.W.W. (International Workers of the World).

BUMPER POLE   The hardwood pole riding between the two sets of runners on a logging sleigh. Same as rooster, goose neck, dray hook, slipper.

BUNCH   To skid logs together at some convenient place for loading

BUNCHING   Binding logs together with a cable when hoisting them to load a sleigh or box car. Same as choking.

BUNCHING CHAIN   1. A log chain for skidding or dragging a log. Same as skidding chain. 2. A skidding chain wherewith more than one log was skidded at one time, in short: "to skid a bunch of logs in one haul".

BUNCH IT   To quit work.

BUNCH LOGS   To collect logs in one place for loading.

BUNCH TEAM   A team used to bunch logs.

BUNDLING   a method of transporting pulpwood down a steep slope. A chain is bound around about one half cord of wood that is then dragged down the slope to a central point.

BUNG   The hole in the water tank where water is let out to sprinkle the ice roads.

BUNK   1. A lumberjack's bed. The early bunks had no springs. They were generally pole shelves with straw or balsam boughs for cushioning, no mattresses. 2. The heavy timber upon which the logs rested on a logging sleigh. 3. The crossbeam on a log car upon which the logs rested. Same as bolster.

BUNK BOUND   An unbalanced load of logs that let the bunk get low on one side when the front runners were turned so it could not be straightened.

BUNK CHAIN   A short chain to tie logs on the sleigh or to extend a longer chain. Same as toggle chain.

BUNK HOOK   The hook attached to the end of the bunk on a logging car, which may be raised to hold the logs in place or lowered to release them.

BUNKHOUSE   The sleeping quarters of the lumberjacks. Same as sleeping shanty.

BUNK LOAD   A road of logs not more than one log deep, i.e., in which every log rests on the bunk.

BUNK LOGS   The outside logs on a logging sleigh chained in position.

— B —

BUNKO   A man hired to steer lumberjacks into "dives" or "bawdy houses".

BUNK SHACK   Same as bunkhouse or men's camp.

BUNK SPIKES   Sharp spikes set upright in the bunks of a logging sleigh to keep the logs in place.

BUNTING BAR   Used as a battering ram to jar loose loaded sleighs or railroad cars wheels which were frozen down during especially cold nights.

BUNTING POLE   A pole between sleds of a sleigh to hold them apart.

BURN AUGER or **BURNER**   An auger heated red hot in order to burn its way into wood.

BURNER   (Waste or refuse burner)—A huge metal furnace at a sawmill used to burn sawdust, wood chips and other wood wastes accumulating at a sawmill.

BURN OUT THE GREASE   Go to town to get drunk.

BUSH   The woods or the back country. The back forty. The word was used more commonly in Canada than in the lake states.

BUSH A ROAD   To mark the route of a logging road in the woods or marsh, or on the ice by setting up bushes.

BUSH RANGER   A cruiser.

BUSH TAIL   A horse.

BUSTAKOGAN or **BUSTADSHOGEN**   In verse, timberland or forest.

BUST THE BOOMS   Too much water and too many logs will break the chains forming the log pond behind the dam.

BUTCHER'S POLL AXE   A single-butted axe having a short (3 inch) blunt spike opposite the cutting edge. Used for killing cattle prior to butchering.

BUTT   The lower part or base of a tree trunk. As verb, to mark off the lower part to be left standing or discarded in the woods because it would not make good lumber. Same as butt cut, butt log.

BUTT CUT, BUTT LOG   1. The first log above the stump. 2. In gathering tanbark, the sections of bark taken from the butt of the tree before felling it for further peeling.

BUTT MARKS   Same as end marks.

BUTTERFLY HOOK   A hook used in the early pine days, curved in such a way as to prevent a line or chain from popping out when slack was given.

BUTTERIS   A chisel used to pare horses' hoofs in shoeing. A farrier's tool used to trim the horse's hoof before the shoes are put on.

BUTTING SAW   Saw used to remove rotted butt log ends at mill.

BUTTING TIMBER   Rounding off log ends to make them skid more easily. Same as ross.

BUTT LASH   The proclivity of a tree, when felled, to pop forward, sideways, or backward off the stump, often very dangerous to the fallers.

BUTT OFF   1. To cut a piece from the end of a log because of a defect. 2. To square the end of a log.

BUTTRESS   A wall or abutment built along a stream to prevent the logs in a drive from cutting the band or jamming.

BUTT ROT   Rotted or decayed end at the base of a tree. This portion was usually cut off.

BUTT TEAM   The team next to the logs. Same as pole team, wheelers.

BUY MY SHEEP   A rough-house lumberjack game played nights in the bunkhouse. The blind-folded jack (usually a "greenhorn") gets his bottom soaked in water.

BY THE BUSHEL, BY THE INCH, BY THE MILE, or BY THE PIECE Contract work done by the piece or unit, but usually applied to contract felling and bucking.

BY THE FOOT   How jacks got to camp or left camp—they walked.

BUZZ SAW   Same as circle saw.

After Supper

In the evening the men usually sat around. Some patched mittens or other clothing, some read wornout novels many times passed around, some played poker, some lay on their bunks; but most of the men just sat and smoked. There was talking and story-telling and occasionally a little singing and mouth organ or fiddle playing. Generally, however, this type of entertainment was reserved for Saturday night and Sunday. If the cook or boss had some news it was given at this time after supper.

CABLE ROAD   An aerial mechanism using a cable running from the forest to the landing with wheels mounted on a carriage affixed to the cable. Logs were "skidded" out of the woods a foot or two off the ground.

CABOOSE   Large cast iron box stoves burning cord wood to keep the bunkhouses warm.

CACKLEBERRIES   Eggs. Same as hen fruit.

CACKLER   A white-collar worker.

CALABOOSE   The lumberjack term for jail.

CALDRON   A large kettle for boiling clothes.

CALIPER   See log caliper.

CALIPER RULE   Scale rule for measuring logs.

CALK or CAULK   1. To drive moss or oakum between logs. 2. The pegs on the soles of a river man's shoes to keep him from slipping off wet logs when on a river drive.

CALKING or CAULKING BAR   A wood or iron bar for driving moss or oakum into the crack between the logs of a building.

CALKING IRON   An iron device with a loose center for driving moss or oakum into the cracks between the logs of a building. Same as calking bar.

CALKS or CAULKS (pronounced "corks" by the rivermen)   1. Short, sharp spikes set in the soles of shoes. River drivers used them to keep from slipping off the logs when they had to ride the logs downstream. Same as corks. 2. Spikes in a horse's shoe.

CAMBOOSE (originally spelled cambus or caboose)   1. The central open fire bed or fireplace directly beneath the smoke hole in the roof in the bunkhouse cook shanty of the very early pine camps. 2. Sometimes, the bunkhouse. 3. On a railroad, the car for the train crew to ride in. 4. The end of the shack with an open-hearth fireplace.

CAMEL BACK   A steel incline to run railroad logging car wheels back on the track after derailment.

CAMP   A logging camp.

CAMPAIGN   French-Canadian term for caboose or open fire in the cook shacks of the very early camps.

CAMP CLERK   Camp bookkeeper.

CAMP DOG   A helper who cared for the bunks and sleeping quarters in a camp.

CAMP FIDDLER   Any jack who played the violin for the entertainment of the other jacks in a camp.

CAMP HUNTER   A hunter who kept the camp supplied with deer meat. There were no game laws in the early logging days.

−C−

CAMP INSPECTOR   A short-time worker, or one who traveled from camp to camp looking for work but refused it when it was offered. Always got a free meal. A lazy lumberjack.

CAMP MORTAR   A mixture of dirt, lime and water to fill the chinks between the logs of the early camp buildings.

CAMP ROBBER   The Canadian jay bird, the whiskey jack. Bird about the size of a robin that stayed around logging operations, especially when fed. Same as lumberjack bird.

CAMP SITE   The location of a logging camp.

CAMP WATCHMAN   A man who guarded the logging camp during the summer months when there was no logging.

CANADA   To use a cant hook to get the logs started up straight when loading on sleigh or cars.

CANADA GRAYS   Warm woolen socks, German socks.

CANADA PEAKER   A system of arranging logs on a load in the form of a triangle. The load was built by placing logs in the notches of the other logs.

CANADA SIDE   Opposite of the face side. Same as gin pole side.

CANADAW   French logger's term for Canada.

CANADIAN WAY   Practice of cutting logs into lengths instead of hauling entire log trunks to the landing.

CANADY   Jack's term for Canada.

CANAYEN   Canadian, as: "He's a Canayen."

CANDYSIDE   That crew of a high land-camp that was best equipped. The other side is naturally the haywire one.

CANNONED   A log being sent up skids to a rollway, or a load when it got out of control and pointed skyward, similar to a cannon.

CANT   1. Refers to hewn timbers or a timber squared in a sawmill. 2. To turn something as with a cant hook.

CANT DOG A riverman's tool. A short-handled peavey. Once described by a greenhorn as a stick with a hook hanging on it. It was used on a drive for dislodging logs from the banks and in breaking jams. A classic remark heard on a drive when a riverman fell in the river was "To hell with the man, save the cant dog". Same as cant hook, mooley.

CANT HOOK A tool like a peavey, but having a toe ring and lip at the end instead of a pike. Used to handle logs on land. Same as crooked steel, mooley, swing dog, cant dog. Sometimes called a log wrench.

CANT HOOK MAN A logger working at a landing or rollway.

CANT HOOK STALK Cant hook handle.

CANT HOOK TIP HEADER Blacksmith-made die into which hot metal is driven to shape the tip ends of a cant hook.

CANTS Logs with two side slabs removed to produce a flat surface in order to run through the gang saws in a sawmill.

CANT SETTER Sawmill workers who set the wood cants on the saw carriage.

CANUCK A Canadian logger.

CAPER Any festivity, but usually a dance.

CAPERING CLOTHES Dress-up or party clothes.

CAPPOT or CAPOTE Hooded overcoat like a parka. A Canadian term.

CAPSTAN A vertical cleated drum revolving on an upright spindle and used in moving very heavy weights.

CARCASS SPREADER A curved piece of wood or iron used by butchers to hold a butchered carcass apart at the hocks for sawing down the backbone to split the entire carcass into two equal halves. Same as gambrel.

CARD LOGS To guide logs through white water downstream to the sawmills.

CARD MAN A union man. Often shortened to card.

—C—

CARRIAGE The moving platform upon which logs were fed against the saws.

CARRIAGE RIDER 1. One who rides the head rig carriage in sawing lumber. Same as dogger. 2. Sawmill workers who rode the moving saw carriage and performed the necessary work involved with the carriage.

CARRY THE BALLOON To search for employment.

CAR STAKES Stakes on the sides of a flatcar to prevent logs from falling off.

CAR STARTER A metal shoe with a heavy, wooden lever-like handle used on the woods railroads to start log cars moving when the wheels were frozen to the tracks on very cold days. See starting bar.

CARTERS Teamsters on the tote or supply roads or carting roads.

CARTING TEAM Team used on the tote or supply roads.

CAR WIRE Wire used to keep car stakes from spreading when car is loaded.

CAT 1. A caterpillar tractor. Any track tractor used in logging is often called a cat. 2. White spruce; also called skunk, bull, cow.

CATAMARAN A small raft carrying a windlass and grapple, used to recover sunken logs. A windlass. Same as monitor, pontoon, sinker boat.

CATCH BOOM A boom fastened across stream to catch and hold floating logs.

CATCHER'S AXE A very narrow axe with a five- foot-long handle, used to cut side marks into logs prior to floating downriver.

CATCHER'S MARK Same as catch mark.

CATCH MARKER A good river driver who rides on floating logs to identify ownership. He carries a long- handled poll axe to mark logs when original marks are not visible.

CATCH MARKS   Marks cut into or stamped upon logs by river crews of log boom companies to denote ownership.

CATCHMENT FACILITIES   All facilities of a boom company such as wing booms, sorting booms, storage booms, etc.

CAT DOCTOR or CAT SKINNER   One who drives a caterpillar tractor in a logging operation.

CAT FACE   A spot on the tree indicating rot or fire scar in the wood.

CATHEDRAL WOODS   A beautiful stand of pine, the trees tall, straight, and thick with a soft carpet of pine needles.

CAT PIECE   A small stick with holes at regular intervals, placed on top of uprights firmly set in floating booms. The uprights were fitted to enter the holes in the cat piece, at the entrance to a sluiceway or sorting jack. The cat piece is held by the uprights high enough above water to allow logs to float freely under it.

CATS PAW   Chain attachment by binding the loose end of a chain to the hook without tying a knot.

CATTLE   Term used for oxen.

CATTLE POWER   Oxen power.

CATTY MAN   Any man who was quick on his feet, generally in reference to a river pig.

CATTY ON HIS CALKS   Same as catty man.

CAT WALKS   Rough plank walks or squared timber walks connecting the various sections of the sorting works on the booming grounds.

C.C.C.   Civilian Conservation Corps.

CEDAR ITCH   An allergic response to cedar pollen making the late spring and summer logging of cedar unpopular to many jacks.

CEDAR LOGGING   Same as cedar pole camp.

CEDAR POLE CAMP   A logging operation that cuts only cedar poles.

CEDAR SALVAGE   One who cuts or peels cedar logs, poles, or posts.

—C—

CEDAR SAW   A very thin cross- cut saw from 6–7 feet in length with no raker teeth.

CEDAR SHAKE SPLITTER   A metal, cone-shaped tool used to split cedar rounds into blocks for making shingles by hand with a shingle froe.

CEDAR SPUD   A flat, chisel-like tool with a long handle (approximately 4–5 feet) used to de-bark cedar posts and poles.

CENTER   A few logs held on a single rock or obstruction in a river during a log drive.

CENTER CHAIN   The middle chain used to hold logs in place on a railroad car when the load was in motion.

CENTER JAM or **CENTRE JAME**   A log jam on an island or other obstruction in the center of the stream. Same as stream jam.

CHAIN   A unit of measure used by a timber cruiser on a land survey. It is sixty-six feet long; five chains equal one tally.

CHAIN BOOM   Any boom in which individual boom sticks are held together by boom chains.

CHAIN DOGS   A skidding device having two chains with sharp dogs at one end of each chain with the other ends attached to a ring.

CHAIN MAN   Same as chainer or chain tender.

CHAINER   A teamster's helper in skidding logs out of the woods.

CHAIN GIRDLE   An intricate log mark resembling a chain which extended approximately half way around the log and put on as a bark mark or side mark using a catcher's axe.

CHAIN GRAPPLES   Skidding tongs.

CHAINING   Skidding logs with horses and chains, as opposed to the use of a go-devil or dray.

CHAIN LOADING   Loading logs with a single chain hitched to horses or oxen rolling them up the skid pole onto the sleigh.

CHAIN SAW   A power-driven saw for falling and bucking with teeth set singly in an endless chain running around an oval tract.

CHAIN TENDER   One who assists in skidding or loading operations. Same as chainer, sled tender.

CHAIN TIGHTENER   A leverlike device for tightening corner binds or wrappers on a load of logs.

CHANCE   A term used to define the ease or difficulty with which a particular logging operation can be conducted. A good change indicates favorable conditions for easy logging. Also called logger's chance.

CHANGE HIS GAIT   "Do differently the next time"—change in thinking.

CHANNEL   Same as flume.

CHASER   1. A drink imbibed after having a "shot" of whiskey. Usually beer. 2. Man who unhooks logs from chocker, high-lead logging.

CHAW   Chewing tobacco.

CHEATER   Nickname for the camp clerk or the scaler.

CHEAT STICK   A scale rule for estimating board feet in logs. Same as log scale rule, scale rule, money maker, robbers cane, thief stick, swindle stick.

CHECK   1. A longitudinal crack in timber caused by too rapid seasoning. Same as season check. 2. One complete course of four logs in a crib.

CHECK SCALER   A company scaler who came to camps to check on the scales being made by the camp scaler.

CHEEK PLATES   Metal plates enclosing a sheath or wheel of a block.

CHICKADEE   A road monkey whose job was to clean the horse manure from the ice road.

CHICKEN FEED   Scrap trimmings or sawmill refuse.

40

—C—

CHIENNING   A French word meaning dog's work, a hand-sled type of logging. Used in eastern Canada in rough terrain.

CHIPPEWAY   Jack's term for the Chippewa River in Wisconsin. At times used to designate Chippewa Falls, Wisconsin.

CHIPPEWAY LIGHTNING   A very cheap whiskey.

CHIPPEWA PINE   Select pine growing in the famous valley of the Chippewa River in Wisconsin.

CHIPPEWA RAFTS   A special type of lumber raft used on the Chippewa River.

CHIN CHOPPER   Tree that splits in falling.

CHINK   To fill with soft clay, oakum, or moss the crevices between logs. Same as moss, mud.

CHIN-WHISKERED JOBBER   A logger operating on a small scale. Same as zippo.

CHIP-IN-A-ZEE   The practice of chopping a notch on the side on which a tree was to fall.

CHIPPER   An axe man or chopper.

CHISEL HOOK   A type of cant hook having a hook with a chisel edge.

CHOCKER or CHOKER   A wire loop or cable used in yarding logs or skidding chains. Also called whiskers.

CHOKER MAN or CHOKER SETTER   One who fastens or sets the choker on the logs in yarding logs.

CHOKE STRAP   A necktie.

CHOKING   Binding logs together with a cable when hoisting them to load a sleigh or railroad car. Same as bunching.

CHOP   To cut down trees, cut up into logs, or cut off limbs.

CHOPPER   1. One who cuts down trees. 2. Jacks who specialized in chopping down trees with an axe without the aid of a saw.

CHOPPING IRON   A axe.

CHOPPINGS   The area where lumberjacks are cutting down trees.

CHORE BOY   One who cleaned up the sleeping quarters and stable in a logging camp; the cook's assistant who did such work as cutting firewood, building fires, and carrying water. Same as bar room man, bull cook, flunky, shanty boy, shanty boss.

CHOW   A meal.

CHUCK   A supply of food for the lumberjacks.

CHUCK BOAT   A wanigan. A boat with supplies that followed the log drive downstream.

CHUCKHOUSE   The cook shacks in a logging camp.

CHUKE   Indian name for the knit caps often worn by the old timber cruisers and land lookers.

CHUNK OUT   To clean skid roads, especially to remove chunks.

CHURN BUTTED   A tree greatly enlarged at the base. Same as bottle butted, swell butted.

CHUTE   A flume for transporting logs. An opening to float timber through or over a dam.

CIGAR BOX VIOLIN   A simple violin made by jacks using a wooden cigar box as the waist of the instrument. Played with a conventional bow.

CINDER HOE   A short-handled, metal hoe used to haul or pull cinders from the fire box of a locomotive.

CINDER SNATCHER   Person who rides on open end of train—or open back end.

CIRCLE HOOK   A large circular-shaped hook on end of skidding chain that loosened easily in skidding logs.

CIRCLE SAW or CIRCULAR SAW   A round saw with teeth all along the circumference and often called a rotary saw.

-C-

CIRCULAR SAWYER   Operator of a circle saw in a sawmill.

CLAM   A machine used in loading and unloading logs.

CLAM GUN   A power shovel. Originally a steam shovel that opened and closed.

CLAW BAR   Heavy steel bar with a claw turned up at the base, used in logging railroad work for pulling spikes. Working end of the bar may be wedge-shaped and used as a pry or a lever. Similar to a crowbar.

CLEAN CUTTING or CLEAR CUTTING   Removing all merchantable timber.

CLEANER TEETH or CLEARING TEETH   The raker teeth on a crosscut saw that pulled out the sawdust, not the cutting teeth.

CLEAR CUT   An area that has been cut over, that is, all merchantable timber has been removed.

CLEARS   Full-fledged members of I.W.W. (International Workers of the World).

CLEAR LUMBER   Top grade lumber.

CLEAR TIMBER   Logs without knots, number one quality.

CLEAR YEAR   Any year when river drives came down with no log jams.

CLEVIS   Stout iron loop used in skidding. Same as shackle.

CLICKING   Pounding on the horse shoe nail head while holding the Dolly beneath the end of the nail so that the nail will bend over drawing the shoe tightly to the hoof.

CLIMAX   1. Name of a locomotive used in logging operations. It was a gear-driven engine. 2. A smoking or chewing tobacco.

CLINCHER   A farrier's tool fashioned like a chisel with a very short handle used to cut off the excess ends of horse shoe nails.

CLIPPER   A plierlike pincer used to bend over and clip off horseshoe nails when shoeing a horse

CLOQUET   A Minnesota "Pine" town. Often called the "home of the white pine". Site of the famous Cloquet forest fire in which Cloquet was "burned out".

CLOSED THE DOOR   Died.

CLOTHES PIN BOLTS   Aspen block from which clothes pins are made.

CLOWN   A city policeman.

CLYDE JAMMER   A railroad type, steam log loader.

COAL CHUTE   A trough made of wood or metal down which coal was transferred from a railroad coal bin to the tender of the loco-motive.

COAL OFF   To cut a forest clean for charcoal wood.

COAL OIL   Kerosene. Also called saw oil. Kerosene was used to light lanterns and torches and to oil the saw in cutting pine so it would not stick from the resin.

COCK SHOP   The camp office.

CODY LOADER   A steam log loader constructed so that railroad flat cars could be pushed beneath it along the railroad track and placed in position for loading. Similar to the McGiffert steam jammer or loader.

COFFEE MILLS   Small, side-wheel steamboats that towed lumber or log rafts.

COLD BEANS   Late for meals.

COLD DECK   A pile of logs left to be loaded and hauled at some later time.

COLD IRON BLACKSMITH   1. Emergency work done away from the camp blacksmith shop; thus, no fire to heat the iron. 2. An inefficient man trying to do blacksmith work.

COLD SHAKE   A discharge; a dismissal.

COLD SHEETS   Fried cakes, doughnuts. Generally, tough doughnuts.

–C–

COLD SHUT or **COLD SHUTS**   1. Usually a repair link in a chain. Can be closed by pounding a pin through a hole on other side of link and burring over to secure in place. 2. Doughnut.

COLLAR   That part of the horse harness fitting over the head and resting against the neck and fore-shoulder.

COME ALONG   A log hook used in skidding logs out of the woods.

COME ALONG CARRY ALL   A grappling hook used by two men for dragging ties or timbers to the place they are needed.

COME AND GET IT   The call to eat, originally the call to noon meal served out of doors because the men were working too far from camp to go in.

COME BACK   Top loader's signal to the cross-haul man to stop horses and return for the next trip.

COME BACK ROAD   Same as go back road.

COME HELL OR HIGH WATER   Without regard for difficulties or consequences. This phrase was coined from river driving and remains in our popular language today.

COMMON LUMBER   Second grade lumber.

COMMON SHINGLES   Shingles of an inferior grade.

COMMISSARY   A small store in a logging camp where supplies are kept. Same as van, wanigan.

COMMISSARY JIMMIE   Same as the camp clerk.

COMPANY BOARDING HOUSE   Many sawmill towns were "company owned". That is; company homes for mill workers, company store and often the company boarding house.

COMPANY HOUSE   A hotel owned by a logging company.

COMPANY MAN   A logger who is overanxious to please the boss. In labor organization work, a worker who has the interests of the employer rather than those of the workers at heart. A foreman or salaried employee. An employee who spies on his fellow workers.

COMPANY MONEY   Script printed by the company and generally accepted in company town except by the U.S. post office.

COMPANY STORE   General store owned and operated by the logging company.

COMPANY TOWN   A community village where the logging company owns all the town except the U.S. post office.

COMPASS SIGHTS   Early timber cruisers and land lookers would sight their pocket compasses (taking a bearing) at a distant mark on the landscape, then walk to that mark, take a second bearing with their pocket compasses and walk to that, thus making progress in a given direction.

CONDENSED DAYLIGHT   A torch or light used to load or unload logs at night.

CONDUCTOR   Teamster's helper on the sprinkler tank.

CONE   A pyramid or mandril. Heavy iron cone used in blacksmith shops in shaping rounds of all kinds.

CONIFER   Any cone-bearing tree—the so-called "evergreens".

CONK   A rotten tree. The decay in the wood of a tree caused by a fungus.

CONK KNOTS (See CONK)   Knots having a fungus-caused decay in them.

CONKY   Affected by conk.

CONMAN   A conductor on a sprinkling tank.

CONTRACT LOGGER   A logger who has a contract to do any part or all of a logging operation.

COOK BOAT   Same as wanigan.

COOK CAMP   A building used as kitchen and dining room in a logging camp. Same as cook house, cook shanty.

—C—

COOKEE or **COOKIE**   Any kind of cook's helper. Same as flunky, slush cook, taffle.

COOK HOUSE   The kitchen and dining room in a logging camp. Same as cook camp, cook shanty.

COOK HOUSE GONG   Same as gut hammer or dinner triangle.

COOKERY BOAT (See **WANIGAN**)   A cook's raft used on river drives.

COOKING BUTTER   Butter that has gone stale in transit or from lack of cooling. When this happened, bacon grease was preferred.

COOKING STOVE   A large cast iron, wood-burning stove (6–8 pot openings), usually a Joesting or Monarch stove, used for all meal preparations.

COOK ROOM   The kitchen and dining room of a camp. Sometimes a separate room where food is prepared.

COOK SHACK   A cook house or cook shanty.

COOK SHANTY   The building used as kitchen and dining room in a logging camp. Same as cook camp, cook house.

COOK TENT   A tent for cooking set up on the bank of a river where the log drive was being made.

COOK WANIGAN   Houseboat used to prepare meals for river drivers. Wanigan.

COOP   The local jail.

COOPER'S ADZE   A short-handled tool for hewing small timbers, barrel staves, barrel hoops, and water yokes.

COOTIE CAGE   A bunk or bed in camp quarters.

COPENHAGEN   Snuff. Same as Scandihoovian dynamite, snoose, Swedish brain food, Swedish conditioner powder.

CORD   Standard cord is a stack of wood four feet high, four feet wide, and eight feet long. It equals 128 cubic feet and contains approximately ninety cubic feet of solid wood and bark.

CORD CUTTER   1. One who cuts pulpwood at a given rate per cord. A piece worker. 2. A hand-made knife for cutting heavy cord or rope.

CORDELLING   Pulling log and lumber rafts with ropes pulled by men on the river banks.

CORDUROY ROAD   A road built by laying poles across a soft spot or swamp. Any road made by laying poles crosswise.

CORDWOOD   1. Fuel wood generally cut in four-foot lengths and used in box stoves to heat the bunkhouse. 2. Fuel wood piled in four-foot lengths for easy measurement. 3. Standing timber only fit for fuel.

CORK PINE   A white pine tree yielding a superior grade of lumber. It floated better than other kinds of pine. Found in Maine and in the lake states.

CORKING HIMSELF   A horse cutting himself with a very sharp shoe point.

CORKS   The calks (short, sharp spikes) set in the soles of shoes of a river driver or in the shoes of horses.

CORKSCREW   A Shay locomotive.

CORNCRIB   A very thin horse in poor physical condition. With ribs showing, it reminded men from the farms of a corncrib.

CORNER   1. In felling timber, to cut through the sapwood on all sides to prevent the trunk from splitting as it falls from its stump. 2. Also refers to a section corner in a land survey.

CORNER BIND HOOK   Used to bind the outside logs of a load to a sleigh.

CORNER BINDS or **CORNER BIZ**   Four stout chains used on logging sleds to bind the two outside logs of the lower tier to the bunks, and thus give a firm bottom to the load.

CORNER HAND CHAIN   Chain on each end of the sleigh bunks to hold the logs in place on a load.

CORNER MAN   In building on a camp or barn of logs, one who notches the logs so they fit closely together and make a square corner.

CORNER POST   A stake driven in the ground to mark sections and quarter-section corners to designate location of land.

CORN STALK VIOLIN   A very small violin-like instrument made by jacks using green pieces of corn stalks about six inches long with a node at each end. A slit was made to expose a thin strip of the outer layer of the stem and supported by a tiny, wood bridge. The instrument was played like a violin.

COUGAR DEN   A term used for the bunkhouse of a logging camp.

COUGAR MILK   Prohibition-era woods liquor.

COUNTER-HEWING   The finish hewing of a square timber before shipment by boat.

COUNTER JUMPER   A clerk.

COUNTRY CLOTH JACKETS   Heavy wool jackets resembling homespun.

COUNT TIES   To get fired or leave camp. The only way out was to walk the ties of the railroad back to town.

COUNTY HOTEL   A county jail.

COUPLER   Pertains to *raft* coupler. Worker who joined Chippewa Lumber Rafts together to make up the famous Mississippi River rafts.

COUPLING DOG   Short chain with dogs at both ends, used in old skiffing operations where logs were fastened end to end.

COUPLING GRAB   Skidding tongs.

COURSE   A single stack of boards, usually all the same size. Several courses piled adjacent and bound by crossers would make a pile with a certain size front.

COVERING   A forest fire in the tops of the trees. Same as crown fire. When a forest fire got in the tops of trees, woodmen would say, "She's covering".

COVES or TIMBER COVES   Canadian term for areas near a river where timber rafts were assembled.

COW   Milk.

COW BELL COUNTRY   A farming area.

CRAB   A small raft bearing a windlass and anchor, used to move log rafts upstream or across a lake.

CRACK JACK   A donkey engine. A stationary steam engine.

CRACK: WHIP SAW   A very dangerous cracking (splitting) of a tree from the butt up while being sawed.

CRADLE KNOLLS   Small knolls or mounds of earth that require grading in the construction of logging roads.

CRAPPER   The woods "comfort station". Same as outhouse.

CRAWLER   Lice.

CRAZY CHAIN   The short chain used to hold up the tongue of a water tank not in use. A water tank or sprinkler has a tongue on both ends.

CRAZY DRAG   A dray used to haul logs from woods to skidway. One end of the log is on the dray and the other end drags on the ground. Same as go-devil, snow snake, travois.

CRAZY WHEEL   1. A cable on a drum to control a sleigh of logs down a steep slope. 2. Same as Barienger brake.

CREAMING   Taking only the best trees in the stand.

CREDIT COUPONS   Pay checks.

CRIB   1. A unit of lumber on a raft when lumber was sent down the river from the mill to market. 2. A raft or boom of logs. 3. A wall or abutment built to prevent damage during a river drive; a barrier for reducing the flow of water downstream but extending only part way from shore and thence upstream so as to form a place for storing logs. 4. House of prostitution. 5. Canadian term for early camps. 6. Heavy log structure built on the ice and filled with stones, then sunk to the bottom with tip projecting to form anchors for booms for sorting or holding logs. Also called jam piers when located above a dam to hold back ice and logs to relieve pressure on the dam.

CRIBBER   A horse who gnaws his feeding crib or anything made of wood.

CRIB DAM   A dam built with cribs of logs and planked on the upstream face.

CRIB LOGS   To surround floating logs with a boom and draw them by a windlass on a raft for towing with a steamboat or tug.

CRIBWORK   Structures built up of crisscrossed logs or timbers to make trestles, open foundations, abutments.

CRICK   Jack's term for creek.

CRITTER   Collective for a work animal in a camp, such as horse, ox, or mule. Domestic animal.

CROOK   A curved bole.

CROOKED KNIFE   Hand-made knives with curved handle and blade used by jacks for whittling. Sometimes called a curved knife.

CROOKED STEEL   A cant hook or a peavey.

CROOKED STICK   Ox yoke.

CROSS CHAIN LINK   The link between the roll and the runner.

CROSS CHAINS   Chains connecting the front and rear sleds of a logging sleigh.

CROSS CUT or CROSSCUT SAW   Saw used in felling and bucking trees. Before rakers were put in saws, only axes were used to cut down timber.

CROSSER   A piece of lumber laid cross ways between courses in a pile of lumber.

CROSS FEED   Levers that make the saw carriage in a sawmill go back and forth.

CROSS HAUL   1. A road built for a team that was pulling the cable or chain used to load logs with a jammer or gin pole. The act of hauling logs a short distance with a cable, usually while loading with a jammer. The cleared space in which a team moved in cross hauling. As a practical joke, green help would be sent to camp to get a cross haul or, if they did not have one, to bring back a sulky neck yoke or perhaps a left-handed monkey wrench. 2. A chain around a log to roll a log up the skids onto the load. Sometimes called parbuckle.

CROSS-HAUL JOCKEY   A teamster driving a cross-haul team.

CROSSING PLANKS   Heavy planks laid down where a logging railroad crossed a local road. These planks enabled road vehicles to cross the railroad tracks more easily.

CROSS TIES   Same as ties or railroad ties.

CROTCH   1. A dray for hauling logs, generally made out of the crotch of a tree. 2. The fork of a tree. 3. A loading line that converges into two lines.

CROTCH GRABS   Two hooks connected by short chains to one ring thus forming a V-chain contraption. Same as log shackle.

CROTCH LINE   A chain device for loading logs onto railroad cars or sleighs, or for decking in the woods. Same as croutch chain.

CROTCH TONGUE   Two pieces of wood, in the form of a V, joining the front and rear sleds of a logging sleigh.

CROUTCH CHAIN   A chain device for loading logs. Same as crotch chain.

CROW BAIT   A lean or old horse that is no longer useful.

CROWN   Topmost umbrella of living branches of a tree.

CROWN FIRE   A forest fire that goes into the tops, or crowns, of the trees. When it does that, the woodsmen say, "She's covering". Same as covering.

CROWN SPLIT   A tree trunk that splits from crown toward butt in falling.

—C—

CROWSNEST   One tree that in falling lodges against another. Same as widow maker.

CRUISE   To estimate the amount and value of standing timber. Same as estimate, value.

CRUISER   One who estimates the value of standing timber. Same as estimator, land looker, valuer.

CRUISER'S AXE   A light, short-handled axe carried by timber cruisers.

CRUISER'S SHEET IRON STOVES   A small, light weight, collapsible stove used for the early timber cruisers.

CRUMB BOSS   A low name for a bull cook.

CRUMB CHASER   A cook's helper in the logging camp.

CRUMBS   Lice. Same as blue jackets, gray backs, livestock, small game in bunks. Sidewheelers. Some lumberjacks claimed that if they lined their bunks with cedar boughs lice would not get on them.

CRUMMY   Ridden with body lice.

CULL   Rejected logs having little or no value.

CULLER   Man who quickly graded and sorted lumber being cut at a mill by picking out the culls or rejects. Same as sorter.

CUSTOM CUT   To cut logs for only one sawmill.

CUSTOM LOGGERS   Small-scale loggers who cut and hauled for the large companies.

CUT   A season's output of logs.

CUT A LOG   1. To move one end of a log forward or backward, so that the log will roll in the desired direction.

CUTAWAY DAM   A temporary dam, built on a tributary to hold back the water. It was usually built of slash and removed after the drive had passed.

CUT 'ER   A command given to let go, when a cant hook man is holding back a log rolling down a skidway.

CUT LUMBER   Lumber from a sawmill.

CUT ME DOWN   A crosscut saw.

CUT OFF   An artificial channel by which the course of a stream is straightened to facilitate log driving.

CUT OFF SAW   A saw used in making fuel wood.

CUT OUT AND GET OUT   Get the logging operation done as soon as possible to avoid any more taxes.

CUTOVER   Any area that has been logged, devoid of trees and possibly burned over.

CUT PLUG   Chewing tobacco. Same as plug tobacco.

CUTTER   1. A lumberjack who cut down the tree. A good cutter bragged that he could set a stake in the ground and then fell the tree squarely on top of it. 2. A feller or faller. 3. A small sleigh used by the camp boss to travel between the camps he supervised.

CUTTING CREWS   Small groups of men working together cutting down trees.

CUTTING TEETH   Teeth in a crosscut saw that do the actual cutting, as opposed to clearing teeth or rakers that merely remove the sawdust.

═D═

No Talking at the Table

An ironclad rule in all camps prohibited talking at the table. With 80 to 120 men eating off tin plates and using tin cups there developed a noise hard to describe. Some called it a 'symphony in tin'. It was a rumbling, but there was still the clank of tin plates. Talking wasted time, and the cook wanted the men out so he could clean up and prepare for the next meal. With talking a meal would last almost an hour; with no talking the men were out in fifteen to twenty minutes. Cookees or flunkies were busy from the time the men sat down elbow to elbow filling and refilling the two-quart dishups. Lumberjacks really could eat. Almost before the men left the table cookies were washing the dishes in boiling hot water. "Silverware" case knives were dried quickly by shaking them back and forth in a grain sack.

DAN PATCH   A cut plug tobacco favored by jacks.

DARK BURLEY   A popular tobacco in the early lumber camps. Usually chewed.

DASH BOARD OVERALLS   Bib overalls.

DAYLIGHT IN THE SWAMP   A call after breakfast to start the men into the woods to work. A bull cook once asked the boss what time he should wake the teamsters. The reply was, "Any damn time you catch them sleeping". Loggers referred to all logging as "letting daylight into the swamp". Same as rollout.

DAYLIGHT ROBBER   A log scaler.

DAY MAN   A laborer who receives a stated daily remuneration.

D.B.H.   Initials for "diameter breast high". It means 4½ feet from the ground, the height at which the diameter of a tree is measured.

DEACON   Master of ceremonies in the bunkhouse.

DEACON SEAT or DEACON BENCH   1. The one classic piece of camp furniture, built on the outer end of muzzle loading bunks. Usually made of half a log, flat side up. The men sat around the fire before turning in, resting, smoking, and talking. There were no store chairs in the early logging camps. 2. Used also as a table for eating when the cook shanty and the sleeping quarters were in one building called the camboose.

DEAD AND DOWN   Standing trees and trees on the ground either dead or living.

DEADENER   A heavy log or timber with spikes set in the butt end, so fastened in a log slide that the logs passing under it came in contact with the spikes and would have their speed retarded.

DEADENING   Same as girdling: consists of cutting a ring around the tree trunk deep enough to stop normal sap circulation in the tree.

DEAD HEAD   1. A water-soaked log lying on the bottom of a river or lake, or a partly sunken log. Same as bobber, sinker. 2. One who idles on the job, very lazy.

DEADHEADING   Salvaging sunken logs.

DEAD HORSE   A sort of useless job. River pigs considered stream clearing dead horse work.

—D—

DEAD MAN   1. A fallen tree on the shore, or a timber to which the hawser of a boom is attached. 2. A log buried in the ground to which a guy line or an anchor line is attached.

DEAD WATER   That part of a stream having such slight fall that no current is apparent. Same as still water.

DEAL   A large, hewed timber with rounded corners for exporting, usually to England.

DEATH WARRANT   A hospital ticket. The lumberjacks' "Blue Cross".

DECK   A pile of logs in the woods, at a landing, or at a mill. Same as log deck.

DECKER   One who rolls logs upon a log deck. He straightens out the logs and arranges them on the pile.

DECKER LOADER   Early steam log loader used with comparatively small logs of the lake states.

DECKING CHAIN   A long chain used in loading or piling logs. Same as loading chain.

DECKING JAMMER   A jammer used only for piling and decking logs.

DECKING LINE   Same as decking chain.

DECKING ROPE   Same as decking line.

DECK UP   To pile up logs at a landing.

DEER FOOT   1. A V-shaped iron catch on the side of a logging car in which the binding chain is fastened. 2. The unique shape of the butt end of one type of single-bit axe handle.

DEHORN   1. To saw off the ends of logs bearing the owner's mark and put on a new mark. 2. Term used by old-time wobblies to denote anything that takes the mind of the worker from the class struggle.

DENNISON HOOK   A hook used in the early pine days in handling logs to keep the line secured.

DENS   Houses of prostitution, cribs.

DENTIST   One who files saws.

DERRICK   The loading boom on railroad log-loading rigs.

DEXTER   Same as sled knee.

DIMENSION STUFF   Lumber cut in specific lengths and sold in carload lots. Same as piece cut.

DINGLE   The roofed-over space between the kitchen and the sleeping quarters in a logging camp, used as a storeroom for meat and wood. Same as alley, alley way.

DINKEY or DINKY   A small logging locomotive.

DINNER HOLE   Where the noon meal is served out in the woods.

DINNER HORN   A conical-shaped tin horn about 4–7 feet long used for calling lumberjacks to meals or, sometimes, for getting them up in the morning. Same as gabriel, gut horn.

DINNERING OUT   Same as nooning.

DINNER TRIANGLE   Heavy iron rod bent into an unclosed triangle that the bull cook or punk boy would ring to call the jacks to eat. It is said it could be heard for a mile or more. Same as gut hammer.

DIPPER   Short for steam shovel mounted on a railroad car.

DIRECTION BOX   A compass.

D-IRON   A clevis-shaped iron mounted on the end of an A-frame woods jammer and upon which the top block is fastened.

DIRT HIDER   A road grader.

DISH UP TABLE   Serving table in the cook shanty.

DISTRICT LUMBER SCALER   A lumber inspector.

DIVE   A rough and tough type saloon. A house of ill fame.

DOCK   1. To cut a man's pay. 2. A long, raised platform from which lumber is piled in a lumber yard.

DOCKING SAW   Same as pit saw or whipsaw.

━D━

DOCK WALLOPERS   Stevedors or longshoremen who loaded lumber on boats.

DOCTOR BALSAM   The balsam pitch lumberjacks often used as a medicine for colds and sore throats or as a salve on cuts or cracks.

DOG   1. A device attached to the runner of a sleigh that digs into the ground and prevents the sleigh from sliding backward when going uphill when the team is rested. 2. A device for holding a log in place when sawing into lumber. 3. A short, heavy piece of steel, bent and pointed at one end to form a hook and with an eye or ring at the other. It was used for many purposes in logging, and is sometimes so shaped that a blow directly against the line of draft will loosen it. Same as tail hook.

DOGGER   One who rides the head rig carriage in a sawmill and sets the dogs into the wood to hold the logs in place while being sawed. Same as carriage rider.

DOG HOOK   Hook coupled with a short chain and used to bind logs together in horse skidding.

DOG IRONS   Used by the bull cook to hold logs in place while sawing wood for the cookstove.

DOG KNOCKER or **DOG HAMMER**   A small sledge used to knock chain hook dogs from logs.

DOG MARK   Mark on the face of the board left by the dogs on the head blocks.

DOGS   1. Skidding tongs. Same as skidding hook. 2. A lumberjack's feet.

DOG WARP   A rope with a strong hook on the end, used in breaking dangerous jams on falls and rapids and in moving logs from other difficult positions.

DOG WEDGE   An iron wedge with a ring in the butt, which is driven into the end of a log and a chain hitched in the ring for skidding the log for horsepower. Also used in gathering up logs on a drive by running a rope through the rings and pulling a number of logs at a time through marshes or partially submerged meadows to the channel.

DOLLAR A DAY   Pay rate for ordinary labor in camps in the 1890s.

DOLLY   1. A two-wheel cart for hauling lumber. Same as buggy, lumber buggy. 2. A metal block used by farriers (blacksmiths) to draw the horse shoe tightly against the hoof by striking with a hammer.

DONKEY   Short for the steam donkey engine used for skidding and loading logs. Same as donkey engine. A stationary engine.

DONKEY DOCTOR   A donkey mechanic.

DONKEY ENGINE   1. A stationary engine. 2. A steam engine equipped with a drive shaft and flanged spool over which cables were run to skid, pile or lift logs in loading.

DONKEY MAN   The engineer who runs the donkey engine.

DONKEY PUNCHER   A man who runs a donkey engine used in logging. A donkey skinner.

DONKEY SKINNER   A man who runs a small engine in the woods. A donkey puncher.

DOOR KNOBS   Biscuits.

DOTE   Decay or rot in timber. Same as dozy.

DOUBLE-BITTED AXE   An axe with two sharpened edges. It was more generally used in the woods than the single-bitted or poll axe.

DOUBLE COUPLER   Two sets of two each of spreader chains fastened to a single ring.

DOUBLE-CUTTING BAND SAW   A band saw having teeth on both edges of the saw thus permitting the saw to cut on both strokes of the carriage.

DOUBLE-DECKED BUNKS   Lumberjacks' beds in the bunkhouse. Made two tiers high. Often called double deckers.

─D─

DOUBLE EDGERS   Edgers with two saws to cut off the outside edges of a board to make it standard width.

DOUBLE HEADER   A place from which it is possible to haul a full load of logs to the landing, and where partial loads are topped out or finished to the full capacity of team.

DOUBLE PINE   Same as buckwheat pine.

DOUBLING   Cutting a train-load of logs in two in order to pull it up a steep grade.

DOUBLE RACK   A sleigh body designed to carry two tiers of four-foot wood.

DOUBLE ROTARY SAW   A saw rig whereby one rotary saw was aligned above the other as the head saw in preparing to feed cants into the gang saw in a mill.

DOUBLETREE   A hardwood crosspiece to which single trees were attached when pulling with two animals in a skidding operation.

DOUGH BOXER   A cook.

DOUGHBOYS   The camp cook and his helpers. Often referred to as river drive cooks.

DOUGH GOD   Camp bread.

DOUGHNUT HOOK   Same as doughnut lifter.

DOUGHNUT LIFTER or DONUT LIFTER   A woven wire screen with which to lift doughnuts from the boiling fat, allowing them to drain. Blacksmith-fashioned.

DOUGH POUNDER   A baker.

DOUGH ROLLER   A cook.

DOWN BELOW   The lower peninsula of Michigan.

DOUSE THE GLIM   Turn out the light! (Kerosene lamps of the logging camps.)

DOWNHILL CLEVIS   A brake on a logging sled consisting of a clevis encircling the runner, to the bottom of which a heavy square piece of iron is welded.

DOWNHILL HAUL   Any woods job which is easy going.

DOWN RIVER   To the sawmills.

DOWN THE HILL   A cry to let other workers know a tree was falling. Same as tim- ber.

DOWN THE LINE   Going to town.

DOWN THE PIKE   A statement indicating that a lumberjack was quitting his job. Same as make her out, hit the pike, hit the trail, mix me a walk, histe the turkey.

DOYLE'S RULE or DOYLE'S SCALE   One of several rules used for scaling timber.

DOZY   Decayed. Same as dote.

DRAFT CHAIN   A chain used to pull load attached to the ox yoke.

DRAG   1. A device for leveling logging roads. 2. Rakers or cleaning teeth on a crosscut saw.

DRAG CART   A small truck with two low wheels used in skidding logs. Same as bummer.

DRAG DAY   That day of the month when a man can draw his wages in advance of the day they are due.

DRAG 'ER   To quit.

DRAGGERS   Derelict, laggard logs out of the main current in a river drive.

DRAG HAULING   Skidding, especially as was done with horses.

DRAG IN   To skid a log. Same as dray in or skid.

DRAG ROAD   A long skid road from woods to skidway. Same as dray road, runway, gutter road, travois road.

—D—

DRAG SAW   Any crosscut saw, portable or stationary, powered by a small gasoline engine. Some early models were powered by steam in Wisconsin.

DRAG SLED   A two-runner sled used to haul logs out of the woods. Same as dray, bob, crotch, drag sled, lizard, scoot, skidding sled, sloop, yarding sled.

DRAINER   Usually a blacksmith-made iron drying rack used in camp cook shanty to hold tin and enamel plates.

DRAUGHT CHAIN   Same as skidding chain.

DRAWBAR   The connecting bar between couplers on log cars.

DRAW DOWN   Permitting excess water to flow through a dam by opening gates.

DRAWING A TREE   Notching a tree to make it fall in a desired direction.

DRAW KNIFE   A wood-worker's tool having a blade with handles at each end, used to shave off surfaces by drawing it toward one.

DRAW SHAVE   Same as draw knife.

DRAY   Two runners with a bunk in the center to haul logs out of the woods; a single sled used in dragging logs. One end of the log rests upon the sled, the other on the ground. Same as bob, crotch, drag sled, lizard, scoot, skidding sled, sloop, yarding sled.

DRAY DAY   That day of the month when a man can draw his wages in advance of the day they are due. Same as drag day.

DRAY-HAUL CAMPS   Camps where drays were used to haul logs short distances to landings or banking grounds.

DRAY HOOK   A bar used to couple logging sleighs. Same as goose neck, rooster, bumper pole, slipper.

DRAY IN   To skid a log. Same as drag in, skid.

DRAYING LOGS   Skidding logs with a go-devil.

DRAY ROAD   A long skid road from woods to skidway. Same as drag road, runway, gutter road, travois road.

DRAY STEP   A skid used to pile logs on a dray.

DRAY WISHBONES   Eight to ten foot long pieces of hard maple with a gradual curve at one end, used as runners for a dray.

DRESSED LUMBER   Lumber which has been surfaced on one or more sides.

DRESSES   Horse's harness.

DRIFTER   A camp jumper who wandered from camp to camp never staying very long in one place. Sometimes called a boomer.

DRIFT PIN   A heavy pointed iron rod used to anchor timbers.

DRILLING MACHINE   A mechanical, hand-operated device used for drilling holes in heavy timbers for the insertion of wooden pegs or metal bars.

DRIVE   The floating of logs on a river from the forest to the mill or shipping point. Same as float.

DRIVER   A man working in a river crew as a river pig.

DRIVING BOOTS   Pegged boots for riding logs on drives.

DRIVING CREW   The vanguard on a log drive whose job was to try to keep the logs floating in mid-stream.

DRIVING HEAD or **DRIVING PITCH**   High water suitable for driving logs down a river.

DRIVE MASTER   An experienced river man appointed by a boom company to supervise all driving activities.

DRIVEWAY   The channel at a sorting works where logs passed through as they were being sorted as to ownership.

DROP CAKE PANS   Muffin tins.

DROP LANDING   A landing in which logs would be decked by hand, such as on a side hill.

−D−

DROP LOG A log used on top of dam gate to raise the head of water.

DR. PERRY DAVIS'S PAINKILLER An early patented medicine popular with the jacks.

DRUM A large metal spool on which the steel cable is wound on a donkey engine.

DRUMMERS Salesmen who frequented the camps, often on Sundays, hawking such needs as liniments, tobaccos, clothing, etc.

DRYASS A sack filled with hay for a teamster to sit on when riding on snowy, frozen logs.

DRY FEET Name given to jacks who never worked on the river drives.

DRYING RACK A blacksmith-made, metal rack used for drying the camp enamel or tin plates in the cook shanty. Same as drainer.

DRY KILL Trees killed by flooding. Often found in areas flooded by beaver dams.

DRY KILN An artificially heated chamber for drying and seasoning newly cut, green lumber.

DRY-KYE Driftwood and dead trees in the water along the edge of a pond or lake. Variously spelled.

DRY PICK As applied to a log jam, to remove logs singly while the water is cut off.

DRY REAR Logs of a river drive left stranded on the sand of the streams which must be dragged or rolled back into the water.

DRY ROLL 1. To roll stranded logs into the water behind the drive. 2. An area where logs had become stranded on dry ground because of a rapid lowering of the water on a spring drive.

DRY ROT Decay in timber without apparent moisture.

DRY SHED A building near the sawmill where lumber was dried or seasoned, usually hardwood.

DRY SLIDE   A trough, usually made of logs, for moving logs down a slope by gravity. Same as slide.

DRY SLOOP   To drag logs on bare ground when the slope is so steep that it would be dangerous to drag on snow.

DUCKBILL HOOK   One of the styles of cant hook and peavey hooks used. Shaped like the bill of a duck.

DUDE   One who starts woods work in street clothes.

DUE BILL   A time slip issued by logging companies.

DUFFLE   The personal belongings of a woodsman or lumberjack which he takes into the woods. Same as dunnage.

DUFFLE BAG   A packsack in which a lumberjack carries his belongings. A turkey.

DUGOUT   A crude Indian boat made by burning out all but an outer shell of a pine log and scraping out the charred wood. Lumberjacks made dugouts by hollowing out a log with an adze; used on log drives before the bateau was made.

DUKE'S MIXTURE   1. Any mixed up situation. 2. An old-time mixture of pipe tobacco. 3. A popular tobacco used to roll your own cigarettes.

DUMP   Same as log dump or landing.

DUMP HOOK   A levered chain grab hook attached to the evener to which a team is hitched in loading logs. Tripping the lever releases the hook from the logging chain without stopping the team.

DUMPING GROUNDS   Same as dump or log dump.

DUMP LOGS   To roll logs over a bluff, or from a logging car or a sled into the water.

DUNGHISTER   A farmer. To call a lumberjack a farmer was an insult and meant a fight.

DUNG SNUFFER   A teamster.

=D=

DUNNAGE   1. The personal belongings of a lumberjack. Same as duffle. 2. Lumber below a merchantable grade.

DUPLEX   A type of donkey loading engine that both yards and loads logs.

DUST A DAM   To fill up with earth or gravel the cracks or small holes between planks in the gate of a splash dam.

DUTCHMAN   1. A splinter or slab of wood to be put under a wrapper chain to keep it tight, or to level up the load on a sleigh. 2. Wood left in undercut to support the heavy side of a tree when it was falling on a swing. 3. A short stick placed traversely between the outer logs of a load to keep any logs from falling off. Same as grouser. 4. A short piece of railroad rail filling in a joint in the track.

DUTCH OVEN   Heavy cast-iron cooking pot with tight-fitting lid. Popular in the old pine camps when the cook used to prepare food over the open fire of the caboose. Same as marmit.

DYNAMITE AUGER   Long shanked auger, five to six feet, used to bore holes into which dynamite charges were placed, usually under stumps and rocks.

A Big Day's Work

A team or pair of sawyers was expected to cut an average of about one hundred logs per day of fair white pine timber. In the best pine country, this number might be doubled. In big pine forests, forty or fifty trees could be a big day's work. One lumberjack claimed he himself had cut one hundred one trees, which made four hundred four logs.

The largest load of pine logs was forty thousand board feet, supposed to have been loaded near Lost Lake, Wisconsin. There are many claims to the largest load of pine logs.

EASY AS FALLING OFF A LOG   Expression said to have originated with the river pigs who knew how easy it was to get wet in cold water.

EATING TENT   Generally a fly tent rather than a complete tent attached to the cook tent where the river drivers ate.

EDGING   Cutting off barked edges of boards which have just passed through the gang saw.

EDGINGS   Scraps cut off a board by the edger.

EDGING SAW   Saws that cut off any barked edges from lumber already cut by other saws in a mill.

—E—

EIGHT BROTHERS   A popular camp tobacco.

EIGHT WHEEL WAGON   A heavy wagon having eight wheels and used in summer logging transportation of logs to landings or rollways.

EMPEROR BOX STOVE   A large cast iron box-shaped stove burning cord wood and used to heat the bunkhouses of the camps.

EMPTIES   Any type of logging vehicle being taken into the woods empty before being loaded.

ENDGATE   Cheese.

END HOOKS   Various types of hooks affixed to each end of logs in loading. Hooks were freed by pulling a guide line, after which the logs dropped in place on the flatcar or other loading vehicle.

END MARK or **END STAMP**   A log mark on the end of a log.

ENGINE HILL   Any steep hill where a hoisting engine was set up to pull sleighloads of logs up the steep grade.

ENGINE ROOM   Building housing the steam generating engine in a sawmill.

ENTER LAND   To purchase land on which a timber estimate has been made by a timber cruiser.

EPSOM SALTS   1. A camp doctor. 2. A physic used in the early camps.

EQUALIZER   A whiffletree carried high on the rumps of the horses and used when loading logs with a cable or chain. It had a long-handled trip hook to carry a single chain.

ESTIMATE   To value the amount of standing timber. Same as cruise, value.

ESTIMATOR   A man who values the amount of standing timber. Same as cruiser, land looker, valuer.

EVENERS   A whiffletree which could be set to give a weaker horse the advantage in pulling a load. A device for equalizing the pull between the horses in a team.

Superstitions

Some lumberjacks would not stay in a bunkhouse if there were poplar (popple) logs in it. It was believed that the cross Christ was crucified on was made of poplar. They thought it would bring them bad luck.

Lumberjacks never ground their axes on Sunday. They thought it was bad luck and that they would probably cut their feet the next time their axes were used.

Lumberjacks would never think of starting a log drive on a Friday, no matter if the water was favorable on that day.

FACE   The side of a load of logs where loading has taken place.

FACE LOG   The front bottom log on a skidway. Same as head log.

FALLER   One who chopped down trees; a sawyer. While both faller and feller have been used, by far the more commonly used word is faller. Writers differ in the spelling. New England and Wisconsin prefer the word faller; in other areas, including Michigan, feller is just as common.

FALLER'S WEDGE   A wedge with a rather narrow angle and having a clasp for a chain.

FALLING OR FELLING   Chopping or sawing down trees and cutting them into logs.

—F—

FALLING AXE   Originally a single-bitted axe which gave way to the double-bitted axe used in chopping down trees.

FALLING WEDGE   A wedge driven in the saw cut to keep saw from pinching or to fall a tree in the right direction. Same as felling wedge.

FANCY WOMEN   Whores.

FAREWELL MAN   A camp welfare man.

FARMER LOGGERS   Farmers who went into the camps in the winter time to augment their farm income.

FARRIER   The man in the camp, usually the blacksmith, who shoed the horses and very often cared for their injuries and ills.

FAT PINE   Pitchy wood often found in the base of the butt log or in the stump.

FEATHERS   Marsh hay used to stuff grain sacks to be used as bunkhouse pillows.

FEED BAG   A nose bag as applied to horses.

FEEDER   A man who cared for the horses and stables in a logging camp. Same as barn boss.

FELLER   A lumberjack who cuts down trees. Same as faller.

FELLING WEDGE   A wedge used to throw a tree in the desired direction by driving it into the kerf. Same as falling wedge.

FELLY or FELLOE   The outside wood portion of a wagon wheel.

FENDER BOOM   A boom on either side of a gate to guide logs through sluice gates. A boom so secured that it guided floating logs in the desired direction. Same as glancing boom, sheer boom.

FENDER SKID   A log placed on the lower side of a skidding trail on a slope to hold the log on the trail while being skidded. Same as breastwork log, glancer, sheer skid.

FHARRIGAN   A place where supplies are kept. A wanigan.

FID HOOK   A slender, flat hook used to keep another hook from slipping on a chain.

FIDDLE   The bunkhouse violin.

FIELD NOTES   Notes made afield and kept by all surveyors and cruisers.

FILER   One who files the crosscut saws in the woods or the saws in the sawmills. Same as saw fitter or saw filer.

FILER'S SHACK   A separate shack in a logging operation where the filer worked. Often located near the cutting area; usually had large windows or a skylight.

FIN   A log fastened underwater to direct the course of logs floating down river or to swing a boom to one side.

FIN BOOM   A boom with fins to direct the course of logs in a stream. Also an adjustable boom used on navigable streams where permanent booms were not allowed.

FINE CUT   A smoking tobacco. (Rarely chewed.)

FINK   1. Anyone who does not carry an I.W.W. (International Workers of the World) red card. 2. A stool pigeon, company guard, private detective. If you don't mean it, you better smile, and smile wide and handsome, when you call a man that.

FINN GRUB HOE or FINN HOE   A heavy hoe made from a shovel with a 4 to 5 foot handle slightly curved downward used for leveling tote roads.

FINN SAW   A short saw with bow-type handle, generally used to cut pulpwood. Same as bow saw, Swede saw.

FINN SCRIBE   A metal divider used by Finlanders to mark logs for fitting into log buildings.

FIR   Short for balsam fir.

FIRE BACK   An area removed from the general area of the sawmill where slabs, edgings, and waste material were burned as a safety precaution against fire.

—F—

FIRE CRACKERS   Beanhole beans.

FIRE HUNTING   Shining deer at night with a reflected light or lantern. It was done from a canoe around the edge of a lake. An illegal method of hunting deer.

FIRE RAKE   An iron bar with a right-angle at one end used to stir the coals at the blacksmith's forge.

FIRE SCAR   Any burn mark on a tree, usually near the base.

FIRE TOWER   A high tower placed upon a high point of land where lookouts watch for forest fires.

FIRST COOK   In a large camp, the pastry cook and the boss.

FISH   Anyone who was good for a touch. A person easily fooled.

FISH EGGS   Tapioca pudding.

FISH PLATE   Stout metal plate spanning two railroad ties. Rail ends were fastened together over this plate, making a joint.

FIT A SAW   To put a saw in good condition by jointing, setting, and filing the teeth.

FIT HOOK   A grab-type hook made from a flat iron used to secure the wrapper chain on a loaded sleigh.

FITTER   One who notched a tree for felling and, after it was felled, marked the log into lengths into which it was to be cut.

FITTING UP CREWS   Crews of men who constructed lumber and log rafts.

FLAGGIN'S   Dinner toted out to the woods on a sled by cookee. Standard fare consisted of kettles of roast beef and brown gravy and a couple of bushels of boiled potatoes and homemade bread.

FLAMBEAU FLARE   Large kerosene flares mounted on the prows of bateau working at night on the Flambeau River in Wisconsin.

FLAMBEAU TORCH   Tin twin-spouted torch resembling a double-spouted teapot. Generally used to light a road or landing.

FLANKERS   Rivermen who worked the river edges to keep all logs floating downriver in the main channel.

FLAP JACKS   Pancakes or griddle cakes. Stovelids, flats.

FLARE   A torch burning lard and kerosene used as light in night work, such as icing roads. Jobber's sun.

FLASH DAM   Usually built on a small tributary to the driving stream which, when opened, gave extra water. Same as squirt dam, splash dam, flood dam.

FLAT   1. A railroad flatcar used to haul logs. 2. Pancake, griddle cake.

FLAT BOOM   A walk made of pieces of square timber held together by bolts and used at a sorting works.

FLAT FEET   Same as dry feet.

FLEA BAG   A cheap flophouse, a louse-ridden hotel, often infested with bedbugs.

FLEET   Several rafts of lumber fastened togther to float down a river with the current.

FLITCH   1. The side of a hog salted and cured; a side of bacon. 2. A piece of lumber with bark on one or both sides.

FLOAT   1. To float logs on a river from the forest to the mill. Same as drive. 2. A very heavy, course file with a long handle used to file off the rough edges of horses' teeth.

FLOAT A HORSE'S TEETH   To file off all rough or broken edges on a horse's teeth so that they would grind food properly.

FLOATAGE   A large mass of logs spread over the surface of a river or a lake.

FLOATER   An itinerant worker; a hobo.

FLOATERS   Cedar logs used as boom sticks to float hardwood logs downstream.

FLOATING OUT FROM UNDER HIS HAT   To fall into the river.

—F—

FLOATING REAR   Logs of a river drive left stranded in still water which may be floated back into the current. May be the same as rear.

FLOATING WANIGANS   Wanigans built on rafts used to feed and lodge river drivers.

FLOATS   Elongated (8–9 inches in length), hollow, wood pieces designed to be threaded into fishing nets to keep the nets afloat. Many wooden ware companies in the lake states turned out millions of these floats.

FLOGGINS   Dinner in the woods.

FLOOD   To drive logs by releasing a head of water confined by a splash dam. Same as splash.

FLOOD DAM   A dam built to store a head of water for driving logs. Same as splash dam, squirt dam, flash dam.

FLOOD GATES   Gates in dams which could be opened and closed to regulate the level of impounded waters.

FLOOD TRASH   Sawmill debris scattered along the river front near a sawmill.

FLOP   A bed.

FLOWAGE RIGHTS   Rights which permitted flooding of lands along the streams by water backed up by the dams and allowing river crews to cross the properties affected, remove logs, and carry out other duties.

FLUME   1. A V-shaped runway of lumber to bring logs out of mountains or steep hills or around falls. Same as sluice, water slide, wet slide. 2. An inclined trough in which water runs, used in transporting logs or timbers. 3. To transport logs or timbers by a flume. Log chute.

FLUMED   Logs being sent down a flume or chute.

FLUME WALKER   The man who keeps logs moving in a flume.

FLUNKY, FLUNKEY or **FLUNKIE**   A chore boy around camp. A bull cook, bar room man, chore boy.

FLUTTERWHEEL   A small water wheel of wood or iron used as a power source for early water sawmills.

FLY BOOM   Boom sticks floating in the middle of a log drive. If a log jam developed below, the boom sticks were fastened to trees on the bank to stop flow of logs.

FLY CAMP   A temporary camp.

FLY COP   Plainclothes policeman.

FLYING DRIVE   A drive, the main portion of which is put through with the utmost dispatch, without stopping to pick the rear.

FLYING PARSON   A minister who occasionally came to camp, usually to ask for a donation to support a community church.

FLY NET   A loosely woven linen or leather string coverlet to keep flies and insects off of horses.

FLY ROLLWAY   A skidway or landing on a steep slope from which the logs are released at once by removing the brace which holds them.

FOG HORN   Smoking pipe.

FORBIDDEN FRUIT   Dried apples, pregnant women.

FORE AND AFT ROAD   A skid road in steep country for sliding logs into water.

FORST LOADER   A gin pole with a swinging boom.

FORTY   Smallest unit of acres in which timber is traded. A subdivision used in timber survey consisting of one-sixteenth of a section. Forty acres of land or timber.

FORTY-FIVE NINETY   Sausage.

FORTY ROD   A very cheap whiskey served in the early logging towns.

FOUND   Board and a bunk to sleep in.

FOUNTAIN   Trade name for a shingle mill.

—F—

FOUR CUTTER   Cordwood. Same as four foot wood.

FOUR FOOT WOOD   Cordwood. Same as four cutter.

FOUR PAWS   Same as double-coupler.

FOUR UP SKINNER   A teamster using four horses teamed up to do skidding or hauling.

FRAME SAW   A saw enclosed in a frame usually for sawing boards by hand.

FRAMED PIT SAW   Same as frame saw.

FRAMING CHISEL   Stout chisel with a long shank used in heavier work.

FREE LOGS   Unmarked logs carried down a river or washed up on the river banks.

FREE TRADERS   Men who roamed the logging areas trying to buy horses and general logging equipment for re-sale.

FRENCHIE   Jack's term for a Frenchman.

FRESH HEAD   A newly-formed column of water gathered behind a dam.

FRESHET   Spring flooding which helped to carry logs millward.

FRICTION-FEED CARRIAGES   Carriages in a sawmill used to move logs back and forth against the saws.

FRISKED   Robbed while drunk. Same as rolled.

FROE or FROW   A tool for splitting shingles or staves from a block of wood.

FROG   1. The junction of two branches of a flume. 2. A timber placed at the mouth of a slide to direct the logs. 3. The junction of railroad tracks in railroad logging. Same as throw out. 4. A French lumberjack.

FROG EGGS   Tapioca.

FROGS   French-Canadian lumberjacks.

FRONT   1. The front edge denoting the direction in which a forest fire is moving. 2. That area of a forest stand fronting a drivable stream. 3. The end of a stack of lumber in the mill yard facing the loading dock.

FROST CRACK   Vertical splitting of a tree trunk caused by extreme cold.

FROST FOOLER   A mackinaw, pea jacket, reefer, jumper.

FROST HEAVES   Large humps developing under railroad rails due to frost which would cause derailments.

FRUIT   Onions.

FULL   Dead drunk.

FULL HEAD   The greatest amount of water what can be impounded above a dam.

FULL SCALE   Measurement of logs in which no reduction is made for defects.

A Lumberjack's Story of His Accident

An Irish lumberjack was brought into a hospital, in the early logging days, with a few broken ribs. The nurse, full of sympathy, asked him how it happened. He replied, "Well sister, I'll tell you how the whole thing happened. You see, I was up in the woods aloading one cold morning, when I was sending a big burly school marm up on fourth tier, and I see she was going to cannon, so I glams into it to cut her back, when the bitch broke and she comes and caves in a couple of my slats."

A lumberjack in another hospital, when asked by the nurse how he got hurt, replied, "The ground loader threw the beads around a pine log. He claimed he had called for a Saint Croix but he gave a Saginaw; she gunned, broke three of my slats and one of my stilts, and also a very fine skid." The nurse said, "I don't understand." His reply was, "I don't either. He must have been yaps."

GABBOON   spittoon or cuspidor. Found only in better bunkhouses.

GABRIEL or **GABERAL**   A tin horn about three feet long used to call lumberjacks to meals. Same as dinner horn, gut horn.

GAFF   1. A tender or sore spot on an animal's body due to an improperly fitted harness. 2. A swelling on a tree trunk. 3. A protuberance or swelling of almost any kind.

GALLUSES   Suspenders to hold up the lumberjack's stag pants. A popular brand was the Silver King galluses. Same as braces.

GAMBREL   A crooked stick or iron used by butchers.

GANDY DANCER or **GANDY HAND**   Pick and shovel man. A road monkey or worker on a railroad track crew.

GANG EDGER   A multiple-bladed edging saw in a sawmill.

GANG HELPERS   The sawmill workers who moved boards cut by the gang saw in a sawmill.

GANG SAW   A series of saws arranged parallel in one saw rig for sawing lumber.

GANG SAWYER   Operator of a gang saw in a sawmill.

GANGWAY   The incline place upon which logs are moved from the water into a sawmill. Same as jack ladder, log jack, log way, slip.

GAP SORTER   Same as sorter.

GAP STICK   The pole placed across the entrance of a sorting jack to close it when not in use.

GARBAGE CAN   A camp with poor accommodations.

GARD-A-HILL   To keep a logging road on a steep decline in condition for sleigh hauling by use of hay, straw, or sand.

GATE BOOM   A boom swung across a stream to permit passage of logs or timber or closed to hold the same. It may be opened to permit the passage of trash. Same as swing boom.

GATE SAW   Saw that saws barked slabs when log first comes off the jack ladder in the mill. Same as slab saw.

—G—

GATE STEM   Upright timbers in the gate of a wooden dam.

GAVEL   A billy club kept behind the saloon bar by the bar tender to "quiet" unruly jacks. Sometimes referred to as a blackjack.

GAZEBO or **GAXOBO**   A member of the fellowship of woods workers.

GEARED LOCOMOTIVES   A locomotive in which the power was transferred from cylinders to driving wheels by some type of gear mechanism, as in the Heisler, the Shay, and the Climax.

GEE AND HAW   Commands used by oxen and horse drivers to indicate to the animal which way to turn. Gee means turn to the right; haw, turn to the left.

GEE THROW   A line used in raising heavy logs.

GEEZLER   Tongs to handle pulpwood. More often called just "son of a bitch".

GERMAN SOCKS   Warm woolen lumberjack socks.

GET-AWAY ROAD or **GET-AWAY**   A preselected, cleared path to safety which a faller takes when a tree he is cutting starts to fall.

GET HARD NOSED   To get angry.

GET HOLT ON   To get ahold of.

GET IN THE HARNESS   Get to work! No loafing!

GET THE PINK SLIP   To be discharged.

GIANT   A popular chewing and smoking tobacco.

GIG   1. To move along the river's edge. 2. A gadget like a wheelbarrow but with a runner instead of wheels.

GIG BACK   1. A river term meaning to go back to the head of a stretch of rapids to run through it again. 2. To go back to the beginning of a haul over an iced road by a short cut that might carry an empty sleigh but not a load. 3. An attachment for gigging back a sawmill carriage. 4. Rivermen going back upstream to pick up logs stranded at the river edges or on sand or mud flats.

GIGGING HOME or GIGGED BACK   Walking back toward camp on the trail. Walking upstream after the drive is over. A river crew walking back to their base camp at night.

GIG TRAIL or GIGGING TRAIL   1. A trail alongside a stream used to follow log drives. Trout fishermen today use these old gig trails to travel the bank of a stream. 2. A walking path along the river bank used by river drive crews going to and from camp.

GILLEARY or GILRAY   A cant hook. A swing dog gilleary.

GILL POKE   A swinging boom used to poke logs off cars at a log dump. Same as gin pole.

GILT-EDGE CULLS   Cull lumber still salable.

GIN POLE   1. A short spar, used instead of a jammer for loading and unloading logs. 2. Any of the three poles of a hoisting gin. 3. A single pole held in a vertical position by guy wires which support a block and tackle used for lifting logs. Same as gill poke.

GIN POLE SIDE   Opposite of face side. Same as Canada side.

GIRL HOUSE   Place of entertainment for the lumberjacks. Saloon with girl inmates. Sporting houses, house of prostitution.

GIRLS   Dance hall whores.

GIRT   An ox was measured by its girt or girth when sale prices for oxen were agreed upon.

GIVE HER SNOOSE   Meaning to increase power, to hurry. A tribute to the potency of snuff, or snoose, used by loggers.

GIVE HER THE GAPP LA DAW   A French Canadian term meaning to increase production.

GLANCER   A log placed on the lower side of a skidding trail on a slope to hold the log on the trail while being skidded. Same as fender skid, breastwork log, sheer skid.

GLANCING BOOM   A boom on either side of a gate to guide logs through the sluice gate. Same as fender boom, sheer boom.

—G—

GLIM The kerosene lamps in a camp.

GLISSE SKIDS Freshly peeled skids on which logs slide instead of roll when being loaded. Same as slip skids.

GLUT Rather large, heavy saw wedges used along with steel wedges to produce lift in falling trees. Also used for splitting rails. Not commonly used in lake states. (Could be round).

GOAD STICK Hickory rod from four to six feet long with a sharp steel point on one end used in driving oxen. Same as prod.

GOARD STICK Same as goad stick.

GO BACK ROAD A road upon which unloaded sleighs can return to the skidways for reloading, without meeting the loaded sleighs enroute to the landing. Same as short road.

GO-DEVIL A sled made from two natural crooks of maple or ironwood with timber bolted across, for hauling logs. The go-devil kept the end of the logs off the ground as they were dragged out of the woods. Same as travois.

GO-DEVIL WISHBONES Hardwood blanks from which the curved runners of a go-devil were made.

GO FROM ME—COME TO ME A pike pole.

GOLDFISH Canned salmon.

GOLDEN TWINS A popular brand of camp tobacco.

GOOD LOOKING HEIFERS Fancy whores.

GOOD ON THE RIVER Refers to the agile rivermen who rode the logs downriver to the sawmills.

GOOSE NECK 1. A hook made of heavy iron used on the front of a sleigh to hold the evener onto the pole. 2. A bar used to couple two logging sleighs. Same as dray hook, rooster, bumper pole, slipper.

GOOSE-WING AXE A broad axe made for hewing and shaped like the wing of a goose.

GORE STICK STEER   When a camp ox broke a leg, it was butchered and eaten by the jacks. It was said, "Only a jack could chew a gore stick steer since it was so tough".

GOT E'ER MADE or GOT HER MADE   Quitting the job. The lumberjack had his stake made. Saved some money.

GOT SET   An expression used by a teamster when he became mired down or got stuck with a sleighload of logs.

GOT THE ROLLERS PUT UNDER HIM   A jack who got "fired".

GO TO THE WOODS   Farmers in areas nearby to logging shows would augment their farm incomes by working in a logging camp during the winter. This was spoken of as "Going into the woods".

GOUGER   A device for cutting ruts in a road for the runners of a sleigh. Same as rutter, groover, swamp angel.

GOVERNMENT SCALER   Timber scalers employed by the United States Government to check log cuts on U.S. Government lands.

GOVERNOR   An employer; a superintendent; a manager; a director.

GRAB   A company store.

GRAB A ROOT   Grab and hold onto the rope or sucker line of a timber raft when going through a heavy rapids. Holding onto the rope kept one from being swept overboard.

GRAB HOOK   A hook into which a chain could be fastened quickly and securely.

GRAB LINK   An iron link in the shape of a key used to hook anything along a chain.

GRAB MAUL   A hammer often made by camp blacksmiths used to pound grabs or couplers into logs.

GRAB RING   The extra large ring on a grab hook.

GRABS   Skidding tongs. Same as coupling grab.

—G—

GRAB SKIPPER   Hammer with a pointed end used to knock out grabs or couplers from logs.

GRADE   1. The quality of a log. 2. An abandoned railroad or truck road.

GRADE MARKS   Marks stamped into a log indicating quality.

GRADER   1. A four-wheeled vehicle carrying a blade for scraping the surface of a road. Sometimes called a dirt hider. 2. A man who grades lumber.

GRADING CREW   1. Men skilled with hoes, picks, and dynamite for road and railroad building. 2. A crew consisting of a lumber grader and one or more lumber handlers.

GRADERS   sawmill workers who selected the various grades of lumber coming off the saws.

GRAIN   Beans.

GRAIN MEASURE   Sheet iron measuring tins designed for use by horse teamsters to give the correct amount of grain to each horse daily.

GRAIN SCOOP   A large, metal shovel used by teamsters to move bulk grain to feed their horses.

GRANDPA   A general superintendent.

GRAPPLES   Skidding tongs.

TRAPPLING IRON or HOOK   A hooked iron used as an anchor or grab in a rafting operation.

GRAVEL   Salt.

GRAVEL A DAM   To cover with gravel or earth the upstream side of the timber work of a dam, to make it watertight.

GRAVEL HILL   Gravel or sand sprinkled on a downgrade road to check speed of loaded sleighs.

GRAVEYARD DUST   Snuff.

GRAY BACKS   Lice. Also known as crumbs, blue jackets, or body lice.

GREAT GUESSING GAME FROM MAINE   Timber cruising or timber estimating.

GREEN CHAIN   A moving chain from which green lumber is sorted.

GREEN GOLD   Timber, primarily white pine timber of the lake states.

GREENHORN   A city man in the woods.

GREEN TURTLE   A very early brand of plug chewing tobacco. One of the jacks' favorites.

GREEN TWIG RECKONING   A reckoning cruiser often carried a green twig and would bend a kink in it every one hundred paces to record his reckoning.

GRINDSTONE CITY   This Michigan city made sandstone grindstones for most of the logging operations in the lake states.

GRIPS   Skidding tongs.

GROANER or GROWLER   An old rampike which had fallen and lodged in a hardwood tree, where it remained for years. The wind blowing through it set up an awful howling noise.

GROOVER   Sleigh having chisellike blades set to cut ruts in roads that logging sleighs will follow. Same as rutter, swamp angel, gouger.

GROUND HOG   Directed the course of the logs with a peavey as they were rolled upon the car.

GROUND HOGS   Tie cutters who live away from a main camp, far back in the woods.

GROUND LEAD   Type of logging, now rare, where a steam donkey yarded logs along the ground.

GROUND LEVEL   A Finn grub hoe.

GROUND LOADER   A member of the crew who attaches the tongs or loading hooks to the logs or guides the logs up the skids. Same as bottom loader, hooker, hooker on, sender, skids.

GROUND MOLES   Same as chainers.

—G—

GROUND PLANE A pick axe or grub hoe.

GROUND RUTTER A heavy steel rutter which cuts ruts in roads preliminary to icing.

GROUSE LADDERS Limby trees, wolf trees.

GROUSER 1. A device which was let down into the sand to hold a tugboat in place while it was drawing in a log boom. 2. A knot, piece of wood, rock, or anything used as chink when loading. Same as dutchman. 3. A six-foot pole with a spike in each end to hold a load of logs on a hill until it could be controlled with a block and chain.

GROWSING A method of slowing down the drift progress of lumber and logs rafts by shoving long poles into the bottom of the river at the edges of the rafts thus creating a "drag" effect to slow down the progress of a raft especially at turns in the river.

GRUB 1. Food or eats. 2. To dig up small trees and brush with a grub hoe in making woods roads.

GRUBBING IT Doing it the "hard way", usually by hand.

GRUB HOE A heavy handled hoe used in place of a shovel in fire fighting and in leveling roads.

GRUB PILE A cook's call to dinner on a log drive.

GRUB PIN A thirty-inch pin used for fastening in a crib lumber being sent down the river to market. Usually made from the root of an elm or an ironwood tree.

GRUBS Same as grub pin.

GRUB STAKE 1. The money a man has for food when he starts shacking. 2. A wooden stake on a lumber raft that, with one or two others, confines the load.

GUARD A HILL To keep a logging road on a steep incline in condition for use.

GUARD ROLLS   Rollers on either side of a band saw to prevent the saw from moving towards or away from the log being sawed.

GUDGEON PIN   A pin that goes through the front end of the sleigh runners and into the roller to which the tongue is attached. A king pin.

–H–

Log Jam

The great log jam of 1884 at Grandfather Falls on the Wisconsin River north of Wausau is considered by most river drivers to have been the worst log jam in the lake states. Grandfather Falls was indeed a bad place, as the river bed dropped one hundred feet in one and three-quarter miles. The water rushed through ledges and bailed over great jagged rocks that stuck from the bed of the river. Only a few boats had ever gone through safely and then only at highest flood time.

The drive that spring had started from Eagle River, and logs kept accumulating as other logs from the tributaries joined those on the main river. Near Tomahawk other great drives of logs joined them from the Tomahawk and Somo rivers. About one hundred million had been cut on the upper Wisconsin River that year, and about two-thirds of them had gone over the falls safely because of the high water of the early spring. Only about thirty-two million got caught in the jam caused by the flood waters receding enough to expose the rocks in the falls. It took a month to break this jam. After attempts to dislodge the key log, Jim Crane of Oshkosh, who had broken many jams on the Wolf River, was sent for. It took him two weeks to dynamite the logs and send them on their way to mills at Wausau, Stevens Point, and Grand Rapids (Wisconsin Rapids).

HACK   A bark mark or log mark. A mark stamped on a log to denote ownership.

HACKIN   A clearing made by girdling or hacking trees in order to kill them.

HACKMATACK   The hemlock or larch tree.

HAIR POUNDER   A horse teamster.

HALFBREED   A type of donkey engine used in yarding logs.

HAMES   That part of a horse's harness resting on the collar.

HAMES COVERS   Heavy, leather coverlets placed over the hames of a horse harness to keep rain and snow from damaging the hames and horse collars.

HAND BARROW   Two strong, light poles held in position by rungs, upon which bark or wood is carried by two men. Same as ranking bar.

HAND CAR   A small hand-powered railroad vehicle. Sometimes called a pumper.

HAND HOLT   To get hold of something with the hand. As a practical joke, new men were often sent in from the woods to the camp blacksmith for a pair of hand holts.

HAND LOGGING   To move timber without the aid of animal or mechanical draft.

—H—

HAND PIKE A piked lever, usually from six to eight feet long, for handling logs piled in the water. A short pike pole. Hand spike.

HAND RING A ring on the swivel hook, grasped to swing the riggin.

HAND SPIKE Any piece of board or pole used as a pry. Hand pike. Same as jam pike.

HAND WINCH Any winch, used to pull boom sticks tightly against the logs while making up a log raft.

HANDYMAN Same as wood butcher.

HANDYMAN'S SHOP Same as the wood butcher shop in a camp.

HANDY ROD A long iron rod with a hook on the end used in the camp kitchen for lifting dishes down from high shelves or in the blacksmith shop for lifting such things as horseshoes off rafters.

HANG AN AXE To fit a handle or a helve into the axe head at the proper angle so it fitted the owner and probably no other man.

HANG THE BOOM To put the boom in place around a bunch of floating logs.

HANG UP 1. A tree felled so that it lodges against another instead of falling to the ground. Same as buckwheat, lodge. 2. As applied to river driving, to discontinue; thus, a drive may be hung up for lack of water or for some other reason. 3. In hauling with a team, to get the load stuck either in the mud or behind a stump.

HARD HAT A later year term referring to the light metal or plastic hats worn by woods workers as a safety precaution.

HARD SCRABBLE FARM A very unproductive farm, sometimes a rocky hillside farm. Lumberjacks often lived on these farms while working in the woods during the winter.

HARDTACK OUTFIT A concern that sets a poor table, so called for the hard and cheap Swedish bread often served in camp.

HARDTAIL A mule.

HARDWOOD   A broad leaf deciduous tree, as opposed to the conifers or softwoods. Same as broad leaf.

HARDY   A blacksmith's chisel with a square shank which fits into a slot in the anvil with its cutting edge up.

HARNESS AWL   A weighted awl used to punch holes in leather harness materials.

HARNESS BULLS   Policemen.

HARNESS HOOKS   Large metal or wood hooks (often made from a tree crotch) used to hang up harnesses at the end of the day's work.

HARPOON   1. A one to two-tined spear used for getting logs out of the water. 2. A spear used for spearing fish through the ice in the early spring or in the water.

HASH CHOPPER   One of the very few mechanical devices or tools in a camp kitchen used to chop up the "makins" of hash and consisted of a rotating drum with a vertical chopping blade with the entire mechanism being driven by a hand crank.

HASH HOUSE   Usually, a cheap restaurant in a logging town.

HASH KNIFE   A heavy, curved blade with a handle at both ends used in a chopping motion rather than in a carving or cutting motion.

HASH SLINGER   A waiter in the logging camp.

HAUL   In logging, the distance and route over which teams traveled between two given points.

HAUL BACK   The line that returns the chokers to the woods after a turn of logs has been pulled in.

HAULING ROAD   Main road leading from woods to landing, or rollways, or banking grounds.

HAWSER   A large rope for towing or mooring rafts of logs.

HAY   Money in a pay envelope.

HAY BURNER   A large logging horse, a heavy eater.

─H─

HAY HILL   A hill on which hay or dirt has been sprinkled to act as a brake on a sleigh runner.

HAY HOOK   A long-handled hook device for pulling hay out of a haystack.

HAY KNIFE   A heavy, short-handled knife used for cutting hay bales or hay stacks in winter.

HAY MAN ON THE HILL   One who tended the roads. Same as blue jay, road monkey.

HAYWARD LIGHTNING   Cheap whiskey.

HAYWIRE   1. The well-known binding wire with a thousand purposes in lumber camps. 2. No good; out of order; poor equipment. 3. Now means broken, busted, crazy, foolish, flimsy, or almost anything you think of that isn't as you'd like it.

HAYWIRE CAMP   Same as haywire outfit.

HAYWIRE OUTFIT or **HAYWIRE SHOW**   A contemptuous term for loggers with poor logging equipment.

HEAD   Amount of water in a stream held by a dam.

HEAD AND HIGH TRIP   Same as cross haul.

HEAD BLOCK   1. The log placed under the front end of the skids in a skidway to raise them to the desired height. 2. A log 12 to 14 feet long placed parallel with a skid road and from 2-8 feet away from it and is used as a prop for skids in a cross-haul operation. 3. A piece of wood upon which a raft oar or sweep rested.

HEAD BOOM   Main boom at a rafting works.

HEAD BUMMER   The boss. Head push, side push, working foreman.

HEAD CHOPPER   In the early days of logging in the lake states the head chopper was the sub-foreman who was the boss of a yarding crew which included two fallers, the swampers, teamster, sled tender, and a skidway man.

HEAD DAM   A storage dam for holding back water until it is needed for driving logs downstream.

HEAD DRIVER   An expert river driver stationed at a point where a jam was feared. Head drivers usually worked in pairs. Same as log watch.

HEADERS   Loading at the skidways.

HEAD GATES   The main water gate in a dam.

HEAD LOADER   A man who selected the logs to be loaded on each railroad car.

HEAD LOG   The front bottom log on a skidway. Same as face log.

HEAD OF WATER   Water released by raising the gates in a dam. Logs would float in the extra amount of water.

HEAD PUSH   The boss of a logging operation. Same as side push, straw boss, working foreman. Boss man on a river drive.

HEADQUARTERS   In logging, the distributing point for supplies, equipment, and men; not usually the executive or administration center.

HEADQUARTERS CAMP   A logging camp complex from which several other camps were serviced and supplied.

HEAD RIG   1. Boss in a logging operation. Same as bully, bull of the woods, head rig, main say, woods boss. 2. The main saw in a sawmill.

HEAD SAW   The first saw that logs meet in a sawmill.

HEAD SAWYER   The man who operated the main saw in a sawmill.

HEAD SORTER   Head man or boss at a sorting gap who called out the log marks to assistants responsible for steering logs into the sorting pens.

HEAD TREE   In skyline yarding, the spar at the donkey engine.

HEADWORKS   1. A platform about four feet wide resting on piling on which a man stood when sorting logs above the sorting gap. 2.

-H-

A stationary windlass (Capstan or spool), operated by a horse or by steam power to pull log booms across sizable bodies of water to the lake heads.

HEAT   1. To a timber cruiser, a forty-five minute walk. 2. Every forty-five paces walked by the early land lookers and timber cruisers was called a "heat".

HEAVER   The fireman on a logging locomotive.

HEAVES   A lung disease of horses that prevented proper breathing.

HEAVY   Describes a man or beast with the heaves, asthma, or generally any breathing difficulty.

HE HAS A LONG-HANDLED AXE   Another way of saying that he is cutting over the line on someone else's property; stealing timber.

HEIFER DUST   Snuff.

HEISLER   Company name for one of the geared-type locomotives used in early logging operations.

HE IS WEARING A WOOD KIMONA   He is dead and in his coffin.

HELL TOWN IN THE PINES   Seney, Michigan.

HE LOGS ON SECTION 37   Timber is being stolen from someone else. There are only 36 sections in a township; therefore, a section 37 is a mythical description. Same as round forty.

HELVE   Wooden handle of a tool, especially an axe.

HEMLOCK BARK   Bark of the hemlock tree used for its tanning by tanneries.

HEMLOCK BARKING SPUD   Perhaps the first of all of the tools used in removing the bark from tree trunks.

HEMLOCK BARK SLEIGH   A single rack sleigh used for hauling hemlock bark.

HEMLOCK PEGS   Pegs of all kinds used in beam construction work.

HEMLOCK SHOW   Years ago when hemlock logs were not wanted, any poor show in scrubby timber was called a hemlock show.

HEMLOCK SPECIAL   A long train of sleighs loaded with logs being pulled by a steam hauler.

HEN FRUIT   Eggs. Same as cackleberries.

HERCULES   Trade name for a brand of dynamite. Dynamite was often called just hercules.

HEW   To shape a log into a timber by using a broad axe.

HEWED LOGS   Logs chopped square with a broad axe.

HEWING DOG   A metal bar with spiked ends to hold log in place while hewing.

HEWING HATCHET   A relatively light, small hand axe used by camp wood butchers in the first step of hewing an item out of wood.

HICKEY   Stick used for tightening brakes on a logging train.

HIGHBALL   A command to go ahead. To hurry on.

HIGHBALL OUTFIT   A large or first-class business, especially in lumbering.

HIGH BANKER   A river driver who avoids dangerous places and always manages to find work on a high bank or place less dangeroous and most always a man who shuns hard work.

HIGH CLIMBER or HIGH RIGGER   The man who tops and prepares a high-lead tree for logging.

HIGH GRADE or HIGH GRADING   Taking only the best pine trees in a logging operation or same as "skimming off the cream".

HIGH JACK   A command to throw up arms.

HIGHLEAD   Modern-type power logging where logs are yarded by means of a block hung on spar tree or steel tower or pole, used very little in lake states. Generally a western term.

━H━

HIGH-LEAD TREE   A tree which has been limbed and topped, and from which blocks and tackle are hung for yarding logs.

HIGH ROLLWAY   Winter log pile at the river's edge.

HIGHWATER PANTS   Trousers rolled up halfway to the knees as worn by loggers and other outdoor workers.

HIGH WHEELS   Same as big wheels.

HIKER   A lumberjack who has quit the job. Same as spiker.

HINKLEY KNEES   Same as rave.

HINKLEY'S BITTERS   A cure for colds and other ailments.

HINKLEY'S BONE LINIMENT FOR MEN AND BEAST   Same as Hinkley's bitters.

HISTE THE TURKEY   To take one's personal belongings and leave camp. Same as down the pike, hit the pike, hit the trail, make her out, mix me a walk.

HIT A GILL POKE   On a log drive, to go ashore on the limb of a tree that overhangs the river.

HIT THE KNOTS   To snore.

HIT THE PIKE   Quitting the job. Same as down the pike, make her out, histe the turkey, hit the trail, mix me a walk.

HIT THE TRAIL   To quit one's job in a logging camp. Same as down the pike, make her out, histe the turkey, hit the pike, mix me a walk.

HITTING THE HIGH PLACES or **HITTING THE HIGH SPOTS**   When a jack leaves camp to blow his stake on a fling in town.

HOBNAIL BOOTS   Same as calked boots.

HODAG   1. A grub hoe or mattock. 2. The mythical animal captured by Gene Shepard of Rhinelander in the 1890s.

HOFFLER   In pine country, a man who fastened a hook during loading operations.

HOG   1. A machine which makes fuel out of lumber slabs or refuse to make steam power for sawmills. 2. A gang saw.

HOGAN BOYS   The snakes; D.T. (delerium tremens) caused by drinking barrel whiskey and not eating. As a rule, three such drunks and the lumberjack died.

HOG FUEL   Sawdust.

HOGGER   Same as hog head.

HOG HEAD   A logging locomotive engineer.

HOGSBACK   A very sharp rise or ridge.

HOGSHEADS   Large, wooden barrel in which such staple foods as sugar and flour were delivered to the logging camps.

HOIST   A log loading tripod.

HOISTING BEAM   A heavy square timber placed above the sluice of a dam and used as a fulcrum for the pry.

HOLDING BOOM   A boom for storing logs. Same as receiving boom, storage boom.

HOLDING DAM   A dam usually constructed of earth on logs to impound water for the spring river drives.

HOLDING GROUND   A place in water where logs or pulpwood were held in a boom and released as needed.

HOLLOW BUTT   1. A hollow log with questionable floating ability because of defects such as knots and rot. 2. A log with an open center due to rot.

HOLT TRACTOR   Early caterpillar tractor used to haul several sleigh loads of logs at one time. Affectionately called the "ten ton Holt" because of its weight.

HOLZAXT or SPLIT AXE   A rare axe of German origin used for splitting rails.

–H–

HOME GUARD   A long-time employee of one company. The opposite of boomer and short staker.

HOOF CHISEL   A heavy steel chisel with a handle used to roughly pare horse hoofs in shoeing.

HOOF KNIFE   A knife with a curved tip used for paring the hoof when shoeing a horse.

HOOFLER   A man who attached the hook in loading operations. A word used generally in pine country.

HOOF REST   A wooden or metal tripod upon which a blacksmith could rest the horse's hoof while shoeing the animal.

HOOK   A general name for an almost endless number of devices used in making fast couplings.

HOOKAROON   A tool used in pulling small timbers out of the water or in loading ties on cars. Same as pickaroon.

HOOKER   1. A member of the loading crew who attached tongs to the logs. Same as bottom loader, ground loader, hooker on, sender. 2. A whore or prostitute. 3. Short for hook tender. The boss of one yarding crew in high-lead country.

HOOKER ON   The man on the ground crew who hooked the chain or tongs onto the log for loading. Same as bottom loader, ground loader, hooker, sender.

HOOKMAN   Same as hooker.

HOOK TENDER   The boss of a yarding crew.

HOOK TENDER'S SPECIAL   A four-fingered helping of snoose mixed with a wad of chewing tobacco, recommended by foremen for keeping loggers on the ball.

HOOP SPLITTER   Workers who split wood for barrel hoops, or tool for splitting hoops.

HOOSIER   1. A greenhorn. 2. To slow or botch work.

HOOSIER UP   To slow work purposely or botch up a job.

333333333

HOOT NANNY   1. A small device used to hold a crosscut saw while sawing a log from underneath. 2. A nickname for almost any implement or gadget whose name you do not know. Also called canepas, canepus, kniepus.

HOOTOWL SHIFT   Work that is done very early in the morning.

HORN   1. The dinner horn. 2. Same as cone, mandrel, or pyramid.

HORN BEAM LEVER   A piece of wood made from an ironwood sapling. When dry, it was very strong and durable. In early logging days, it was used considerably before the cant hook was perfected.

HORN BLOWER   The cookee who most always blew the dinner horn.

HORRORS   Visions which resulted from drinking too much whiskey.

HORSE   Same as schnitzelbank or shingle bench.

HORSE BLANKETS   1. Heavy blankets placed over a horse which was heated up from pulling so that he did not get cold and chilled when work stopped. 2. Cigarette papers.

HORSE DAM   A temporary dam made by placing large logs across a stream in order to raise the water level behind it, so as to float the rear.

HORSE HEADWORKS   A large pine log raft with a capstan driven by two horses used to move log booms across large lakes.

HORSE HOVEL   The horse barn in the logging camp.

HORSE JAMMER   A light gin pole strung back to the sled runners and used for loading rather light logs with horses.

HORSE LOGS   Used in river driving to drag or roll stranded logs back to the stream by the use of a peavey. Generally done by two men with cant hook or peavey on each side of a log to lift and pull it where it would float.

HORSE PADS   Heavy pads, usually reinforced cotton batting, placed under horse collars and other pieces of harness to prevent chafing or rubbing.

100

—H—

HORSES' AILMENTS   Heaves, wind broke, spavins, sweneyed, scratches, sore feet, corns, knee sprung, stifled.

HORSESHOE END   A horseshoe hung on a loading skid pole on which logs were rested in a one-man loading operation.

HORSE SHOEING STANCHION   A shoeing stanchion especially made to shoe balky horses by lifting the horse off balance so that the animal could not kick.

HORSESHOE NAILS   Special nails used to hold the metal horse shoe to the hoof of a horse. Were made in two types as: "City Head" and "Regular Head", as well as in various lengths.

HORSESHOE SHAPER   A large steel cone standing about four feet high on which to shape horseshoes. Same as cone.

HORSE SKINNER   A teamster.

HORSE TAIL WHISK   A brush-like switch made from a horse's tail used to keep flies off a horse while shoeing the animal.

HORSE TETHER   Heavy, iron blocks attached to a rope and then to the horse's hames to keep an idle horse from wandering when not working.

HORSE TROUGH STOVE   A small, portable stove placed in the horse trough (drinking trough) in winter to keep the drinking water from freezing.

HORSE WHEEL LOGGING   Skidding logs with big wheels using horses or oxen.

HOSE HOUSE   A small fire station at a sawmill usually housing a portable fire hose on a reel supported by wheels so that the contraption could be hauled quickly to any part of the sawmill works.

HOSPITAL AGENT   A man or sometimes a woman from a hospital who visited camps selling hospital tickets. The lumberjacks' Blue Cross.

HOSPITAL TICKETS or **HOSPITAL CERTIFICATES**   Tickets sold to loggers, usually for ten dollars, entitling the holder to hospitalization and care for a year in the event of injury or sickness. This system was the Blue Cross of early logging days.

HOT BACK or HOT ASS   A roughhouse game played by lumberjacks in the bunkhouse at night.

HOT BISCUITS   Logs.

HOT BOX   Term used to describe a heated bearing on a railroad car, often caused by lack of grease.

HOT DECK   Logs which were to be hauled immediately to the mill, not stored in a pile.

HOT GRAVEL   Dry gravel spread over a grade in a road to serve as a break.

HOT LOGGING   1. A logging operation in which logs went immediately from the woods to the mill. Same as stump logging. 2. Logging involving a spring cut.

HOT POND   Pond at the sawmill kept from freezing by running exhaust steam into it from the engine. Used for holding logs about to be sawed into lumber.

HOT SEAT   Same as hot ass.

HOT SKIDWAYS   Area where logs are transferred from the skidway to the hauling sleds.

HOT STUFF   1. Logs which are not to be decked but hauled as soon as sawed down. 2. Pepper.

HOUSE OF HESITATION   A jail.

HOVEL   A stable for logging horses or oxen. Also a group of camp buildings.

HOVELLED   To bunk with another lumberjack.

HUDSON'S BAY COMPANY BLANKETS   Heavy but warm blankets used by the early timber cruisers and land lookers.

HUNG UP   Refers to a log jam.

—H—

HUNG UP DRIVES Same as hung up. (Often due to lack of spring freshets.)

HUNG TREE A tree which strikes another tree as it is felled, lodging against that tree creating a dangerous situation until it can be felled to the ground. Same as widow maker.

HUNKS Foreign-born laborers, usually unskilled, especially Hungarians or southern Slavs.

HURLEY, HAYWARD AND HELL A very old description of the lumbering towns of Hayward and Hurley, Wisconsin wherein "anything went" as the rough and tumble jacks "shot their wads".

HURLEY LIGHTNING Cheap whiskey.

HURLING DOWN THE PINE Cutting down the pine.

Big Boss Fix

One time a rather large number of men of Hungarian and Polish descent, who had worked in eastern coal mines, were hired, lived with their own nationality, and did not speak or understand any English. One of them, who had never seen an axe, had a tree almost chopped down with a grub hoe before the boss discovered what he was doing. They seemed to have a lot of confidence in the boss; whatever he did was right, and there was nothing he could not do. So, when one of them—a strapping six-footer— accidentally cut off his big toe with an unlucky blow from his axe, he refused hospitalization, although he had a ticket. He kept shaking his head, and saying, "Big boss fix. Big boss fix." So the boss assembled boiling water, carbolic acid for a disinfectant, white silk buttonhole thread, and buckskin needle, and without any anesthetic for the victim sewed the toe back on. He tied the foot to a shingle as a splint or cast, and in a surprisingly short time the man was back to work.

ICE A ROAD   To sprinkle water on a logging road so that a coating of ice would form in which ruts were cut for sleigh runners in hauling logs. A logger reported that an ice road at Rib Lake was used as late as May 15 one year with fifteen inches of ice still intact.

⬛I⬛

ICE AUGER A hand-operated auger used for drilling holes in ice over a pond, lake or river so that water could be obtained in winter to fill the icing sleighs for road icing.

ICE BOX A large watertight box built on a sleigh. It sprinkled the logging roads with water to make an ice road. Same as sprinkler, icer, ice tank, tank, water box, water tank, water wagon.

ICE BREAKERS Ice guards, deflectors to protect bridges and piers during log drives.

ICE GUARDS Heavy timbers fastened fan-shaped around a cluster of boom piles at an angle of approximately thirty degrees to the surface of the water. They prevented the destruction of the boom by ice, by forcing the ice to mount the guards and be broken.

ICER A large watertight box built on a sleigh. It sprinkled the logging roads with water to make an ice road. Same as ice tank, sprinkler, ice box, tank, water box, water tank, water wagon.

ICE-IN TO ICE-OUT The log-driving season on the lake states rivers.

ICING SLEIGH BARREL Hardwood barrels used to fill the tanks of the icing sleds.

ICE RUN Floating ice chunks moving downriver.

ICE TANK A large watertight box built on a sleigh. It sprinkled the logging roads with water to make an ice road. Same as icer, sprinkler, ice box, tank, water box, water tank, water wagon.

I'LL GET THE HAIR OF THE DOG THAT BIT ME The threat of a vanquished man. A vow of a jack who got beaten up in a fist fight.

IMPROVEMENT COMPANY A company in charge of building dams and keeping them in repair to supply water for log drives.

INCLINE A portion of a logging railroad, the grade of which is too steep for the operation of locomotives, and up or down which the log cars are raised or lowered by means of a cable and power. When logs are hauled up grade the incline is sometimes called a hoist.

INDIAN SILKS   Overalls.

INK SLINGER   A logging camp timekeeper.

IN THE BARK   Log rafts sent down the large rivers with bark on the logs—not lumber rafts.

IN THE BRUSH   Out in the woods; working in the timber.

IN-THE-SWING   When oxen are worked in 3–5 teams, the 2nd and 3rd pairs are said to be "in-the-swing" position.

INSPECTORS   Hobos.

IRISH BABY BUGGY   A wheelbarrow.

IRISH MEERSCHAUM   A clay pipe.

IRISH SNOWSTORM   Snow shoveled onto the bare spots of logging roads where the log sleighs passed so that the last of the logs could be plucked from the woods during spring breakup. (Also known as Norwegian or Swedish snowstorm depending on who was doing the shoveling.)

IRON   1. Collective term for all kinds of metal used to make almost anything around a logging camp operation. Included rods, plates, bolts, angle irons, etc. 2. Same as logmarking hammer or stamp hammer.

IRON BURNER   The camp blacksmith

IRON DUKE   Term used for the old iron cookstove of the cook shanty.

IRON MEN   Blacksmiths.

ISLANDS OF PINE   Same as pockets of pine

I.W.W.   Industrial Workers of the World. Paraphrased by opponents as "I won't work! "

–J–

Camp Games

RIDING THE MULE: Two men would hold a cant hook stalk between them. A lumberjack stood on it to see how long he could balance himself before falling off.

ROOSTER FIGHT: A broom handle was placed under the knees of a squatting man with his hands tied to it on either side. Two men tied in this fashion butted each other to see who would get tipped over first.

JACK IN THE DARK WHERE ARE YOU?: A robust, rough-house game played in the bunkhouse. Two jacks are blindfolded; one is "jack in the dark," and the other jack is searching for him. On the query "Jack in the dark where are you?" both jacks would attempt to swat each other with long stockings stuffed with a couple of pair of water-soaked socks. Real swatting resulted.

HOT BACK or HOT ASS: A roughhouse game played by lumberjacks in the bunkhouse. Jacks formed a circle; one blindfolded man was in the center of the circle on his hands and knees. Any jack could swat him on the rear end. The blindfolded man had to guess who had swatted him. If he guessed correctly the swatter became the blindfolded jack in the middle of the circle.

SQUIRREL: One jack would sit down on the deacon seat, with his legs apart and his open hands resting on his knees, palms inward. Another jack, wearing a cap, knelt down on the rough board floor facing him and simultaneously made a noise like a squirrel and ducked his head down to the floor. The sitting man tried to knock off the cap as it passed between his knees. Then the "squirrel" repeated the noise and bobbed his head upward between the other's knees. As one may guess, there were many evasive tricks to this game and many "squirrels" with red cheeks.

J-HOOK   A chisel-edged hook shaped like the letter "J".

JACK   1. A lifting device made of ash or hickory and used to elevate heavy sleighs and wagons. 2. Lumberjack. 3. Money.

JACK CHAIN   An endless chain used for bringing logs from the hot pond into the sawmill. Same as bull chain.

JACK IN THE DARK WHERE ARE YOU?   A robust, roughhouse game played in the bunkhouse.

JACK KNIFE DRAY   A dray loaded so that logs would balance. The driver would step on the back of the logs going downhill to act as brake and step on the front of the load going uphill to make it easier to handle.

JACK LADDER   Heavy, endless chain with dogs which carry logs out of the mill pond up to the gate saws. Same as gangway, log jack, log way, slip.

JACK PINE BARRENS   Rather high, sand country with the larger vegetation being pure strands of Jack Pine.

JACK POT   1. A contemptuous expression applied to an unskillful piece of work in logging. 2. A bad slash. 3. Lodging of more than one tree in felling.

JACKS   Cull pieces of lumber rafts used in prying hung up lumber rafts off mud and sand bars.

-J-

JACK SAW   A folding, hand operated drag saw.

JACK SCREW   A large heavy railroad jack of the screw type.

JACK SLIP   Same as jack ladder.

JACOB'S STAFF   1. A heavy bar attached to the axle of big wheels to raise logs. 2. A staff used to support a staff compass. A Johnson board.

JAG   Being drunk.

JAGGER   A sliver from a wire rope or cable.

JAIL HOOK   Similar to butterfly hook; a large hook, partially closed, used to keep the chain or line from slipping out when slack came into the line.

JAMAICA GINGER   A medicine using ginger as an aromatic stimulant praised by jacks.

JAM   Logs hung up by an obstruction during a river drive. Same as log jam, plug, side jam.

JAM BOATS   Same as bateau.

JAM BOOM   Any boom hastily set up behind the start of a log jam preventing more logs from creating a larger jam.

JAM CRACKER   A riverman expert in breaking log jams.

JAM CREW   The crew who walked down the gig trail alongside the floating logs to keep them from jamming.

JAMMER   1. A derrick for loading logs onto a sleigh or railroad car. Operated by horse or steam. Same as arsneau. 2. Sometimes, a river driver who watched for and helped break up log jams.

JAMMER HEAD   A large, heavy ring of iron with two smaller rings attached to it and used as attachment centers at the tip of a jammer head.

JAMMER PUPS   Same as loading hooks, or "Pups".

JAMMING LOGS   Loading logs on any vehicle, such as a railroad flatcar.

JAM PIKE   A heavy pike with an eight-foot handle and an eight-inch spike at the front end. Used in the early white pine camps by river men. It was supplanted by the peavey. Similar to a pike pole.

JAVA   Coffee.

JAYHAWKING   Stealing stranded logs along the river bank during a log drive.

JEFFERSON BOOTS   A popular brand of calked boots worn by many river drivers.

JEMMY JOHN   The handcar on a railroad track. A pede or velocipede.

JERK THE HASH   Serve the food.

JERK WATER   Coffee.

JERK WIRE   A wire attached to the whistle on a yarding donkey, on which the whistle punk blew signals for starting and stopping.

JESUS BILLY   The evangelist Billy Sunday.

JETHRO IRON   A large metal piece used on railroad ties invented by a man named Jethro.

JEW'S HARP   A ring used to hold the draft chain on an ox yoke.

JIBBO   To remove a hook from a log.

JIGGER   To pull a log by horsepower over a level place in a slide. Same as lazy haul.

JILL POKE or JIL POKE   1. A log stuck in the bank of a stream where it could cause a jam. 2. An awkward person. 3. A wood prop to hold the sleigh in place while loading logs.

JILL-POKER   A lazy fellow lumberjack.

JIM BINDER   A springy pole used for tightening a chain binding together a load of logs. Same as binder.

110

-J-

JIM CROW   A type of rail bender used for bending or straightening steel rails.

JINGLE BELLS   Signal bells: One set in the cab of a steam log hauler and one set in the steersman's cab to signal back and forth.

JINNIE   A drag used to haul logs from woods to skidway. Same as travois, crazy drag, go-devil, snow snake.

JIPPO TRACK LAYERS   A crew hired by a small contractor to lay railroad rails at a given price per foot.

JOBBER   1. A small logging contractor. 2. A rafting pilot.

JOBBER'S CAMP   A logging camp owned and operated by a small contractor or jobber cutting and delivering logs to larger companies.

JOBBER'S SUN   1. The moon; so called in a jobber's or contractor's logging camp because of the early and late hours of commencing and ending work. 2. Kerosene torch used at night so that the jacks could put in longer hours.

JOB HOG   A worker who took another man's position away from him; a man who was very eager to get work.

JOB SHARK   An employment agent who sold jobs to lumberjacks. Same as man catcher.

JOB SIMPLE   1. Fear of losing one's position. 2. Stupid.

JOCKEY CLUB   One of the perfumes preferred by prostitutes.

JOE HOG   A sled for skidding logs.

JOE MOUFFREAU   A legendary Bunyanesque character who roamed the logging camps of the famous Chippewa River Valley in Wisconsin.

JOESTING STOVE   The famous cooking range of the logging camps made in Minneapolis, Minnesota. Had 6 to 8 lids.

JOHN IS UP THE RIVER   A river drive expression indicating that logs will soon be coming down the river.

JOHNNY INKSLINGER   The camp clerk.

JOHNNY NEWCOME   A newcomer.

JOHNSON BOARD   1. A Jacob's staff. 2. A heavy bar attached to the axle of big wheels to help raise logs.

JOHNSON'S LINIMENT   One of the jack's favorite "healalls".

JOINT PINE   Same as cork pine.

JOSEPH PEAVEY   See "Peavey".

JOURNEY or JOURNEY CAKE   A corn pone bread baked in a reflector oven. A johnny cake.

JUDY COON   A raft with a capstan for floating logs across the lake.

JUICERY   A saloon.

JUMP A LOG   Placing a log to fit into a certain place on the load.

JUMBO DRAY   A very large dray used in skidding.

JUMBO SLED or A JUMBO   A modification of the famous go-devil used in skidding runs over swampy land.

JUMPER   1. A sled shod with wood, used for hauling supplies over bare ground into a logging camp. Same as tote sled. 2. A plaid wool jacket. A mackinaw, reefer, pea jacket, frost fooler.

JUMPER SLED   A log sled having a high crosspiece on which one end of a log was supported when it was dragged out of the woods or along a tote road. Used to haul hemlock bark.

JUMPING JACK   A logger who rested all week and worked on Sunday.

JUNE BUG   A lantern set on the booms while loading logs at night.

JUNK WAGON   Same as nooning sled.

-K-

Lumber Rafting

Before railroads were built into all parts of the states, much rough lumber was floated on rafts down the Wisconsin, Chippewa, and St. Croix rivers and then down the Mississippi to the river cities, generally no further than St. Louis. Next it was shipped by rail to the rapidly developing West and Midwest. It was the tremendous demand for lumber in settling the heartland of America that caused such rapid depletion of the pinery.

The smallest unit of a raft is a crib, generally sixteen feet square. Six cribs placed end to end made a rapids piece, and three rapids pieces coupled side by side were known as a Wisconsin raft. A group of Wisconsin rafts were joined together into one large Mississippi raft, generally 116 feet wide and 600 feet long. On the tributaries, the rapids pieces were tied up at night but on the Mississippi they kept moving day and night.

Most of the work was hard, and the raftsmen were always wet from going over falls. An oar, mainly used as a rudder, was 35 feet long, and holding it was a real he-man's job. The fleet crew were experienced rivermen, as it would be unthinkable to lose a lumber raft.

The last lumber raft on the Wisconsin floated down in 1903, and the last raft on the Chippewa went in 1901.

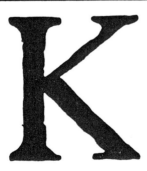

KASO   A tall pail made from birch bark with a small top and broad square base.

KATYDID A set of big wheels, ten to fourteen feet high, used for transporting logs. Same as big wheels, logging wheels, sulky, timber wheels.

KEEPER A boltlike device used to hold the bow of an ox yoke in place.

KEEPING FOOD ON SKIDS TOO LONG Stale food, especially bread and bakery goods that were kept in storage too long.

KEEP THE LOGS MOVING From the woods to the mills, the cry was: "Keep the logs moving!"

KEGGING UP Getting drunk.

KEISTER A packsack, same as knapsack, duffle bag, keenebecker, tussock.

KELLY HAND-MADE AXE A famous, double-bitted axe manufactured by the Kelly Hardware Company.

KENNEBECKER A knapsack, or packsack—generally a carpet bag. It was mostly an eastern term.

KENTUCKS People imported to Forest County, Wisconsin, from Kentucky. They settled throughout the woods and were a great help in logging.

KENTUCKY GIN PANTS Jeans pants made from jean material and often worn in the woods by the early farmer loggers.

KERF The slit or opening made by a saw or axe. The width of a saw cut. Same as saw kerf.

KEROSENE BOTTLE Usually an empty whiskey bottle with grooves cut into the sides of the wood work. Kerosene was sprinkled from the bottle each time a saw cut was made to dissolve the gums and resins that would make the cross cut saw bind.

KEY LOG A log lodged in rocks or caught in some way so that it caused a jam. Its removal would break the jam.

KICKING The action of a falling tree springing back off the stump as it comes down. Dangerous to sawyers.

—K—

KICKS Same as calked boots.

KILHIG Same as Sampson or killing pole.

KILL DAD An empty tin pail where all lumberjacks threw old and odd pieces of chewing or smoking tobacco. Anyone could borrow from it for his pipe.

KILLIG POLE A pole or axe man carried one to push against a tree to make it fall in the right direction.

KILL-ME-QUICKS Big wheels used in summer logging operations. So called because they were hard on horses.

KING LOG The "key" log in a log jam on a river.

KING PIN or KING BOLT 1. The large-headed, heavy pin which passed through the center of the bunk and the beam of the front bob of logging sleds to allow the front, or swinging bunk, to turn at curves in the road. 2. The boss, woods boss, main say, bully, bull of the woods, head rig.

KING SNIPE The boss of a railroad track laying gang. Boss of the steel gang.

KIT All of the gear of a river drive gang.

KITCHEN MECHANIC A dishwasher in a logging camp. Cookee.

KITCHEN SWEAT A dance.

KLONDIKE A famous "boarding house" with painted ladies in the Manistique, Michigan, area. Many girl houses were called Klondike.

KNOCKDOWN BLOCK A block with a movable face plate which could be opened up to lay in a line, thus eliminating threading the line into the block. Same as open side block.

KNOCK HIS EARS DOWN To whip or thrash a person.

KNOT BUMPER A man who cut the knots off logs before they were loaded on truck or car at the landing. Same as knotter.

KNOT SAW   A high speed circle saw used to cut off knots along the edges of shingles.

KNOT SAWING   Cutting off knotty edges in shingle-making.

KNOTCHER or NOTCHER   A lumberjack who notched the tree for falling and cut off limbs.

KNOTTER   A knot bumper.

KNOTTING   Smoothing the branch stubs close to the trunk of a fallen tree.

KNOTTING AXE   Same as swamping axe.

KNURL   A live protuberance, like a huge wart, that grew on some hardwood trees. It might be a foot or two in diameter and project six inches out from the foot of a tree. Because the grain ran in every direction, one made an excellent head for a large wooden maul because it would not split like straight-grain wood.

Wages

Wages varied with economic conditions, but a lumberjack was generally paid from $26 to $30 a month, with board and bunk. The river drivers were paid more, about $50 a month, and the cooks were paid from $80 to $100. A team of horses was worth about the same as a lumberjack.

LACING HOOK   A small metal hook used to re-weave linen horse fly nets.

L HOOKS   End hooks used in loading logs.

LAID OFF   To lay off a felled tree trunk to best advantage in standard lengths and to then saw into logs.

LAKER   A log driver expert at handling logs on a lake.

LAKE STATES   All states bordering on the Great Lakes but the big lumbering states were those on Lakes Michigan and Superior.

LAND HUNTERS   Same as land looker.

LANDING   1. A place to which logs were hauled or skidded preparatory to transportation by water, rail, or truck. Same as bank, banking ground, log dump, rollway, yard. 2. A platform where logs were collected and loaded on cars. See bank.

LANDING CAMPS   Small camps, often near towns used mostly for work in moving logs.

LANDING CANT HOOK   A conventional cant hook but with a very short handle about three feet in length.

LANDING MAN   1. One who unloaded logging sleighs at the landing. 2. Straw boss who kept track of logs arriving at landing and reported same at camp each night.

LAND LOOKER   A man who estimates standing timber. Same as cruiser, estimator, valuer.

LAND SHARK   An unscrupulous land seller.

LANK INSIDE   Hungry.

LAP   Slash left in woods after logging.

LARD   Butter or oleo. Same as salve.

LARD OIL TORCHES or LARD OIL FLARES   Used at night by sprinkler teamsters to make sure of the turns in the iced roads.

LARRIGAN PIE   Same as shoe pack pie.

LARRIGANS   An oil-tanned moccasin with legs used by lumbermen. Both lake states and Canadian lumberjacks wore them.

LASH POLE   A cross pole which held logs together in a raft.

LAST AXE   A small, broad axe used to cut the rough hard maple blocks which were sent to pattern makers in cities to make shoe last models.

LATH   Thin, narrow strips of wood nailed to rafters, ceiling joints, etc., making a groundwork for plastering.

LATH BOLT   A slab or edging cut to correct length, from which lathes are sawed.

LATHERS   River drivers, so-called because they always had a pike pole in their hands which from a distance looked like a lath, hence, the name "Lather".

-L-

LATH SAW   A mill saw used only for sawing lath stock.

LAZY HAUL   To pull a log by horsepower over a level place in a slide. Same as jigger.

LEAD TEAM   1. The front team where two or more teams were hitched to a sleigh. 2. The first team over the road in the morning; usually the best team in camp.

LEADERS   Oxen are worked in teams of from 3–5 yoke. In a team of 5 yoke, the front pair are called the leaders.

LEFT-HANDED MONKEY WRENCH   An imaginary tool used to fool greenhorns in a logging camp.

LEFT TO THE WORMS   Snow covered some logs which were not skidded out of the woods and were said to be "left to the worms".

LICE   Two types plagued the jacks: body lice and crab lice.

LICKDOB   A boot grease used by the early loggers containing lamp-black, tallow and beeswax.

LIFE PRESERVER   Peavey.

LIFE SAVERS   hospital tickets.

LIFT GATE   In a logging dam sluiceway, a gate which may be moved up or down in vertical slides or grooves, fastened to the sides of the sluiceway.

LIGHT BURLEY   A popular chewing tobacco.

LIGHTNING LANDING   A landing having such an incline that the logs may roll upon the cars without too much assistance.

LIGHT SIDE   The top side of a log when flotating in water.

LIMA   A gear-driven locomotive especially suitable for work on heavy grades. sharp curves, and uneven tracks of logging railraods. Same as shay. Lima Locomotive Company called it shay; loggers called it limey.

LIMA CONSOLIDATION.   A locomotive used in lake states logging operations.

LIMA TOW CHAIN   The shay or lima steam locomotive always had a very heavy tow chain with which to tow disabled railroad vehicles.

LIMB   To remove the limbs from a felled tree.

LIMBER   One who cut the limbs from felled trees.

LIMBING OUT   Cutting off the limbs from a fallen tree.

LINE   A rope, as: "Throw me a line!" in river pig's terms.

LINER   The jack who removed the bark from the log and drew a line by snapping a chalk line against it in the process of hewing timbers.

LINING BAR   A heavy steel bar used as a lever for lining up railroad tracks.

LINK AND PIN COUPLING or COUPLER   Early type of couplers used on logging trains. They were very dangerous to the hands and arms of brakemen and other train crew members. An early type of coupling used to hitch one railroad car to another, consisting of a stout pin that fitted into a heavy link attaching one car to another.

LINN HAULER   A latter year, gasolene powered caterpillar tractor used for hauling heavy sled loads of logs over long distances.

LIVESTOCK   Body lice, crumbs, blue jackets, gray backs.

LIVE ROLLS   Heavy, power-driven rollers in a sawmill which carried sawed lumber from the head saw to the resaws.

LIVING ON THE TOWN   A ward of the town.

LIZARD   A sled for skidding logs. A dray, a crotch.

LOADER   1. One who loads logs on sleighs or cars. 2. A steam jammer.

LOADING BAR   A device used in place of doubletrees; a wood bar that hung on the horse's harness to which a long-handled grab hook was attached. Used on cross-haul team in decking or loading logs.

— L —

LOADING BLOCK   A block hung on booms or guy lines in loading operations.

LOADING BOOM   Heavy boom used to swing logs from landing area to cars.

LOADING CHAIN   A long chain used in loading or piling logs with horses. Same as decking chain.

LOADING CREW   jacks who piled the logs on the sleighs.

LOADING HOOK   Any of many different types of hooks used in a loading operation, such as the various jammer hooks. Same as pig's foot jammer hook.

LOADING JACK   A platformed framework upon which logs were hoisted from the water for loading onto cars.

LOADING LINE   A cable used in loading logs on cars with a jammer.

LOADING MACHINE   Steam-powered loading machines mounted on railroad cars, e.g., McGiffert loader.

LOADING TONGS   Large, heavy tongs used in a loading operation.

LOADING TRIPOD   Three long timbers joined at their tops in the shape of a tripod for holding a pulley block in proper position to load logs on cars from a lake, stream, or deck.

LOADS   The number of loaded cars in a logging train.

LOCK DOWN   A strip of tough wood with holes, laid across a raft of logs. Rafting pins driven through the holes into the outside logs held the raft together.

LOCIE   Later day short for locomotive.

LOCO   Short for locomotive. Same as locie.

LODGE   A tree felled so that it lodges against another tree instead of falling to the ground. Same as buckwheat, hung up.

LOG   To cut timber into logs and deliver them to a place from which they can be transported by water or rail to the mill.

LOGAN   A pocket or bay into which logs may float during a drive. Same as poke logan.

LOG BOAT   A long, low, sled-like vehicle used in skidding in wet places.

LOG BOOM   Logs fastened together to hold logs in a stream.

LOG BOOMERS   Same as lumberjacks.

LOG BRAND   Ownership mark stamped on a log with stamping hammer. Same as log mark, side mark.

LOG BRANDING IRON   A branding iron similar to a cattle-branding iron with embossed identification marks used rarely for branding logs.

LOG CHUTE   A flume to transport logs down steep hills or around waterfalls.

LOG DECK   1. A pile of logs in the woods, at a landing, or at the mill. Same as deck. 2. A platform in the sawmill where logs were held until they passed into the headworks of the mill. 3. An even pile of logs made up in the woods or at a landing.

LOG DRIVE   Transporting logs from woods to sawmill down a stream or river. Logs that would float were the pines and most other soft-woods. Hemlock, a softwood, and basswood, a hardwood, would float. All other hardwoods would not float and were not driven down the stream. Tamarack, considered a softwood, would float only well enough to be driven on short drives. Same as drive.

LOG DUMP   A place where logs are hauled preparatory to transporting by water, rail, or truck. Same as banking ground, bank, rollway yard.

LOGGER   One who was engaged in logging. More often referred to the man directly in charge of a logging operation, a contractor or subcontractor.

LOGGERS   Logging horses.

LOGGER'S ROUGH ESTIMATE RULE   Approximately 200 board feet to the log or, five logs to the thousand.

—L—

LOGGER'S RULE   Same as scale stick.

LOGGER'S SMALLPOX   Scars on a lumberjack inflicted by the opponent's spiked boots in a fight.

LOGGING BERRIES   Prunes. A cook was once heard to remark, "For me, I'll take the prune; it makes even better apple pies than the peach."

LOGGING CAMP   A shelter for workers in the woods, generally log buildings. Home or headquarters for the lumberjacks. A temporary structure, it was abandoned when nearby timber was cut. The camp referred to several buildings: the bunkhouse, cook shack, stables, and blacksmith shop. First built of logs with shake roofs, later built of lumber covered with tar paper. Same as camp, set.

LOGGING CAR   Any logging car used in a railroad logging operation, e.g., the Russell car.

LOGGING HORSES   Heavy horses used in logging. Many weighed one ton. They were generally purchased from Iowa or Illinois farms. Also called old plugs, karrons, rocking horses. The average life of horses on sleigh haul and skidding was three to five years.

LOGGING KITS   All of the tools and equipment of a logging jobber.

LOGGING ON THE DEACON SEAT   Tall tales about logging told and re-told by jacks after supper while sitting and smoking on the deacon seat in the bunkhouse.

LOG LIENS   A charge against logs illegally cut on land not owned by the cutter.

LOGGING ON THE RISE   Getting logs into the water in time to ride downstream with the freshet.

LOGGING RAILROAD   Any railroad line on which logs were transported out of the woods to the mills. After the timber was cut away from the rivers, railroads were often used to transport logs.

LOGGING SHOW   A logging operation.

LOGGING SLED or LOGGING SLEIGH   Ordinarily called sleigh in the lake states. Used for hauling logs on snow and ice. Same as bobsled, sled, twin sleds, two sleds.

LOGGING SLED (or SLEIGH) ROAD   A road leading from the woods to the log landing.

LOGGING WHEELS   Big wheels, from ten to fourteen feet high, used for transporting logs from the woods. Same as katydid, sulky, timber wheels, big wheels, high wheels.

LOG GRADES   Quality standards in logs.

LOG HAULER   A steam or gasolene powered engine with a special traction device which is used in place of horses to haul logging sleds.

LOG JACK   An incline plane upon which logs were moved from water to sawmill. Same as gangway, jack ladder, logway, slip.

LOG JAM   Logs caught on an obstruction and piled up during a river drive. It was one of the real problems of early logging. Many river drivers lost their lives in breaking a log jam. Same as jam, plug, side jam.

LOG JAMMER   Any log loading rig.

LOG MAKER   Same as bucker.

LOG MARK   A symbol stamped in the end of a log with a stamping hammer or chopped into the side of the log with an axe to indicate its ownership. Same as log brand, side mark. Used when logs of more than one firm were sent down the river on the same log drive. These log marks were previously selected and recorded in the office of the inspector of each lumber district.

LOG MARKERS   Same as log stamper.

LOG MARKING HAMMER   Same as log stamper.

LOG PENS   Individual log ownership sections at a sorting works.

124

═L═

LOG POOL   Logging companies on the Chippewa River in Wisconsin formed a boom company to handle all the driven logs on the Chippewa River thus saving money for each logging company. The pooled logs were distributed as per their log marks.

LOG RAFTING   Logs lashed together in rafts to be floated downriver to the sawmills.

LOG ROAD   Same as sled road or sleigh road.

LOG ROLLING   An authentic word coming directly from the woods and the drive. A sport where two men vied with each other to stay on a rolling log in the water. A term also used in politics.

LOG SCALE   The board feet content of a log or of a number of logs considered collectively.

LOG SCALE RULE   A long hardwood ruler used in estimating the board feet of lumber in logs or standing timber. Same as cheat stick, scale rule, money maker, robber's cane, thief stick, swindle stick.

LOG SCHACKLE   Same as crotch grabs.

LOG SHOOT   Chute or same as spillway.

LOG SLIDE   A V-shaped trough built down a slope for the purpose of sliding logs to a landing.

LOG SORTER   Same as mark caller.

LOG STAMPER   A stamping hammer, branding hammer, log marker.

LOG STAMPS   Same as log stamper.

LOG, TO   To cut logs and deliver them to a place from which they can be transported by water or rail to the mill.

LOG WAGON   Very heavy four-wheeled wagon for hauling logs, usually used in the summer.

LOG WATCH   An expert river driver stationed at a point where a jam was feared during a river drive. Same as head driver.

125

LOG WAY   The incline plane upon which logs were moved from the water into the sawmill. Same as jack ladder, log jack, slip, gangway.

LOG WORKS   Same as sorting gap.

LOG WRENCH   A term sometimes used for a cant hook.

LOKEY MAN   A man who worked on a locomotive that hauled logs from the woods to the mill.

LOMBARD STEAM HAULER   A crawler steam tractor used in pulling many sleighloads of logs coupled together, made by A. O. Lombard at Waterville, Maine in 1899.

LONG BUTT   The first log of a tree culled for rot or other defects. Generally discarded in the woods as it had no value.

LONG GREENS   Bills in the pay envelopes.

LONG-HANDLED AXES   In the early days of logging it was customary to cut as far over the line as one could reach with an axe, hence long-handled axes.

LONG-HANDLED UNDERWEAR   Heavy wool underwear with long sleeves and legs. Very often they were red in color.

LONG HAUL   The opposite of short haul. Generally a non-profitable haul.

LONG HOG   Sow belly.

LONG JOHNS   Wool, two-piece heavy underwear worn by lumberjacks.

LONG LOGGER   A logger in the fir and redwood country of the West Coast. So called because logs are cut as long as forty feet.

LONG NECK   A bottle of whiskey.

LONG OAT   The horse whip used by teamsters on tote roads and long haul roads.

LONG PINE   Same as cork pine.

LONG ROUTE   To remain at work in one camp for a long period.

—L—

LONG STOVE Same as box stove.

LONG SWEET'NIN Black molasses to sweeten coffee.

LOOKER Same as land looker or cruiser.

LOOSE HOOK TONGS Skidding tongs which have the hooks loosely attached to the ring.

LOOSE LOGS Marked or unmarked logs that had for some reason escaped being rafted. Also called scrabble, prize logs.

LOOSE-TONGUED SLOOP A single sled with wood shod runners and a tongue, used in hauling logs down steep slopes on bare ground. Same as swing dingle.

LOP To cut the limbs from a fallen tree.

LOT Piece of standing timber, small in area.

LOUSE CAGE 1. A hat. 2. Sometimes applied to a bunkhouse.

LOUSY CAMP A logging camp infested with body lice and head lice.

LOWER THE BOOM 1. To fire a logger. 2. Dropping the boom in a loading operation.

LOW WATER BRINGS RUSTY SAWS Logs will never get downriver to the sawmills if the rivers are low.

LUBBER LIFT To raise the end of a log by means of a pry, and through the use of weight instead of strength.

LUCIFER A match. Same as salucifer.

LUFT A block and line attached to the end of the loading line to give more power.

LUG HOOKS A pair of tongs attached to the middle of a short bar and used by two men to carry small logs or railroad crossties. Same as timber carrier, timber grapple, timber hook.

LUMBER To log or to manufacture logs into lumber, or both.

LUMBER A TRACT To cut the trees on a certain specified area.

LUMBER BUGGY   A two-wheel truck used for hauling lumber at a sawmill. Same as buggy, dolly.

LUMBERER   Very early name for the shanty boys.

LUMBER HOOKERS   Sailing vessels which hauled pine lumber from the mills to the cities along the Great Lakes before the advent of the steamships or railroads.

LUMBERING PARTY   Early term for a small camp of men cutting timber.

LUMBERJACK   1. One who worked in a logging operation. The early lumberjacks always lived in a logging camp. Same as shanty man, timber beast, wood hick, shanty boy. 2. A lumberjack is a patient man; he counts to ten and then lets you have it. Sometimes called jacks. 3. A device used to stack high piles of lumber. A metal bar about five feet long, very heavy, used to balance long, heavy planks by a man on ground to a man on top of pile. It was stuck into the pile to anchor it. Thus, by pulling down on one end of a board (plank) leaned against the lumberjack, the other end could be raised to a point where the man on the pile could reach it. 4. A small school girl's description of a lumberjack: "A lumberjack is a man full of bedbugs and fleas. He carries a pack on his back and cuts trees."

LUMBERJACK BIRD   The Canadian jay bird, the whiskey jack. Same as camp robber.

LUMBER GRADING   Classifyying lumber as to quality.

LUMBER KING   A sawmill owner who made much money in lumber.

LUMBERMAN   One engaged in lumbering. Generally referred to one who processes timber rather than one who logs it.

LUMBERMEN'S THREAD   Haywire.

LUMBER PINE   Same as cork pine.

LUMBER RAFT   A collection of lumber fastened together so that it would float. A crib of lumber was 16 by 16 feet and 24 coarses or layers of two-inch lumber. Seven cribs made one rapids piece; three rapids pieces made one Wisconsin raft. A rapids piece was generally 120 feet long.

LUMBER REEFED   Sails on lumber hookers were "reefed up" to accommodate a deck load of lumber.

LUMBER ROLLER   A metal roller device placed at the door of a box car over which pieces of lumber are rolled into the car for transportation.

LUMBER SHOVERS   Men who unloaded lumber from the lumber hookers or a man or boy who handed the lumber to the raftsmen in building a lumber raft.

LUMBER TOWN   A sawmill town.

LUMBERYARD   Piling and storage areas at a sawmill works.

LUNCH   One-third of a day on a river drive, usually about ten o'clock. The time between breakfast at five o'clock and dinner.

LUNCH CAN   A large round can flattened on one side, carried on the back of a lunch carrier or cookee, to carry lunch to the river drivers. It was divided into compartments and would hold lunch for fifty men or more.

LUNCH CARRIERS   Cook's helpers or cookees who carried hot lunch to the lumberjacks in the woods or to the river pigs ahead of the wanigan.

LUNCH GROUNDS or **LUNCH PLACE**   Where a meal was served out of doors, located in a cutting area, usually had a warming fire.

LUNCH HOOK   the hand of the lumberjack.

LUNCH IN   A noon meal served in the dining quarters of the logging camp.

LUNCH OUT   A noon meal carried to the workmen in the woods, either by themselves or by the cookee.

LUNCH SACK A square sack about sixteen inches square usually made of oil cloth with straps to sling over shoulders for river pigs as they worked on the drive. Same as river pig nose bag, nose bag.

LUNG POWDER Snuff.

LUSHERS Jacks who would drink any alcoholic beverage.

LYE Extremely strong tea.

–M–

Plumbing

There was no indoor plumbing in a logging camp. Two poles behind a clump of hemlock or balsam, or in a rough enclosure to keep out snow but not cold, served in lieu of steam-heated facilities. The two poles, one to sit on and one to lean against, kept you in proper position. You then walked or ran back to the camp in the dark, washed up, often in cold water, but felt full of vim and vigor and believed you could lick your weight in wildcats.

M   Symbol for 1,000 board feet of lumber, logs, or standing timber.

MACARONI   Sawdust in long shreds.

MACKINAW   The standard overcoat of a lumberjack. A short wool plaid coat. Legend tells us that in the early days of Upper Michigan (the Mackinaw area), winter coats ordered for the soldiers in that area did not arrive. They did, however, have very heavy blankets that were cut up to make coats later called mackinaws. Same as frost fooler, jumper, pea jacket, reefer.

MACKINAW BOATS   Flat bottom boat with pointed bow and square stem using oars or sails to haul lumber, supplies, and men. Used on the Great Lakes and its tributaries.

MAIL POUCH   A popular chewing tobacco.

MAIN ROADS   Roads used to deliver logs to the river banks.

MAIN SAY or MAIN SPRING   The boss of a logging crew. Same as bull of the woods, bully, woods boss, head rig, king pin.

MAKE HER OUT   What a lumberjack told the timekeeper when he wanted his pay check. A statement made indicating he was quitting. Same as down the pike, hit the pike, histe the turkey, hit the trail, mix me a walk.

MAKENS   Cigarette papers and tobacco. Also spelled makins.

MAKE UP A HAT   Take up a collection for some worthy cause.

MALLET   A short, heavy piece of log on a handle, used to wallop the loaded sleigh runners to free them from the ice after standing overnight.

MALONES   Heavy, wool pants favored by jacks because of the long nap which would shed water and snow readily.

MAN CATCHER   A man employed by a lumber company to recruit lumberjacks. No charge was made, and often the company paid their fares to the camp. An employment agent.

MANDRIL or MANDREL   Heavy iron cone used in the blacksmith shop in shaping rounds of all kinds. Same as cone, pyramid.

MAN THAT OAR   To guide a lumber raft, particularly after it had gone through a heavy falls.

MAN YOKE   A wooden yoke fitting over a man's back at the shoulders used for carrying two pails of water or other supplies.

MARK   A symbol stamped in the end or side of a log with a stamping hammer to indicate ownership. Same as log mark.

MARK CALLER   In sorting logs, one who stood at the lower end of the sorting jack and called the different marks, so that the logs could be guided into the proper channels or pockets.

MARKER   1. One who put the mark on the end or side of logs with a stamping hammer. 2. Worker who notched felled trees where the transverse cuts were to be made by the buckers. The "Marker" was

–M–

sometimes called a swamper who marked the saw cut places as he limbed a felled tree. Same as notcher or undercutter.

MARKING AXE   1. For marking logs for identification. Same as marking iron, marking axe or stamping hammer. 2. A pole axe with log-identifying characters embossed on one edge of the axe.

MARKING HAMMER   A swamping hammer to mark logs for identification. Same as marking iron or marking axe.

MARKING IRON   A bar with raised letters on end for marking logs. Same as stamping iron.

MARMIT   Large cast-iron pot in which food was prepared. A stew kettle. Same as dutch oven.

MARSHALL WELLS HAND MADE AXE   A very popular two-bitted axe made by the Marshall Wells Company.

MARSH FEATHERS   Marsh hay used to cushion the sleeping bunk in the logging camp.

MASTER LUMBERER   An experienced lumberer.

MATE   To place together in a raft of logs of similar size.

MATCHWOOD   Wood from which match sticks are cut.

MATTOCK   Heavy digging tool with a stout hoe blade on one side and an axlike blade on the other. Same as pick axe.

MAUL   Large wood sledge (sometimes metal) bound with metal rings and used for driving heavy stakes or posts.

MAUL PICKAROON   A pickaroon with the opposite edge being a maul.

MCCLAREN CASTINGS   A ball and socket arrangement bolted to the sleigh beam on the runner to give the runner up and down motion or rock.

MCLEAN BOOM   A type of loading boom widely used on the West Coast.

MCGIFFERT LOADER   A railroad steam log loader with booms at both ends. Russell cars could be pulled through the machine itself, which enabled the loader to work from a single track.

MEASURING STICK   Measuring sticks are carried by log makers (buckers) and are marked at four intervals. Same as scale stick.

MEAT BLOCK   A heavy, solid wood block used for a meat carving and meat chopping and sawing block. In camps often a 3 foot section from the trunk of a hardwood tree.

MEAT HOOK   Heavy iron hooks used to hang fresh meat carcasses until ready for use in the camp kitchens. This meat was generally hung in the camp dingle.

MEAT HOUSE   A small, wooden frame building with the door completely covered with wire screen to keep out the flies and usually built in a shady spot to insure coolness to store meat.

MEN CATCHERS   Labor recruiters from the lumber companies who would gather up men in the cities and pay their railroad fare to the logging camp.

MEN'S CAMP   The building in a logging camp that housed and slept the lumberjacks.

MEN SHANTY   Same as bunkhouse.

MERCHANTABLE   That portion of a timber stand or a tree that can profitably be logged and marketed under existing economic conditions.

MERCHANTABLE LOG   A log that will make lumber of a quality and in sufficient amount to make it profitable to take it to a mill and have it sawed.

MERCHANTABLE TIMBER   Usually interpreted to mean timber that can be manufactured at not less than cost. The purpose for which the timber is to be used and local customs are factors which influence the degree of utilization.

MESS PORK   Poorer pork cuts used in camps. Never hams and loins. A very poor cut would be sow belly.

134

-M-

MEXICAN POWDER   Pepper.

MICHIGAN AXE   First of the two-bitted axes. A very popular double-bitted axe originating in the State of Michigan during the early logging days.

MICHIGAN JUMPER   A lumberjack who worked only a few days in one place and then moved on to another camp. His feet got too hot in any one place.

MICHIGAN STAKE   A roll of dollar bills.

MICK or MICKS   An Irish lumberjack.

MIDWAY   The roofed-over space between the bunkhouse and the cook shack where barreled beef, pork, venison, lard, and other supplies were stored.

MILITARY BOUNTY WARRANTS   Warrants that could be assigned to any purchase of pine land, making valuable pine land very cheap at such bargain prices (government land was cheap to begin with—about $1.25 per acre).

MILLING IN-TRANSIT   Pertaining to an Agreement between a railroad company and a lumberman in which the railroad agreed to haul the logs from the woods for a very modest charge, and in turn the lumberman agreed to ship to market over the same railroad a large percentage of all of the lumber it sawed from the logs.

MILLIONAIRE TEAMSTER   Fictional character of many lumberjack tales and tall stories.

MILL POND   The pond near a sawmill in which logs are held until ready to be sawed into lumber. Steam piped from the sawmill kept it from freezing.

MILL RACE   A watercourse directed to move a water wheel that furnished power to a sawmill.

MILL REFUSE   Scrap wood from the sawing operations in a sawmill often used in the burner or in the dry kilns.

MILL RUN   All of the lumber output of a sawmill which had a sale value.

MILL SCALE   The scale of logs delivered to the mill.

MILL STORE   A lumber company-owned store that often paid in time checks or merchandise.

MILL WHEELS   Large sandstone wheels weighing approximately 5,000 pounds used in paper mills to grind wood chips into paper pulp.

MINE OFF   Clear cut timber and move on.

MINNESOTY   Term for Minnesota.

MINUTES   Field notes made by timber cruisers and land lookers.

MIRAMICHI   (Pronounced "mare-ma-she") It was a term commonly used to designate a lumberjack from the Mirimichi River country in New Brunswick, Canada. Here in the lake states there are hundreds of descendants of those loggers.

MISERY WHIP   A crosscut saw used to cut down trees.

MISSISSIPPI RAFTS   A special type of lumber raft used on the Mississippi River.

MIX ME A WALK   Request for time when quitting. Same as down the pike, hit the pike, hit the trail, histe the turkey, make her out.

MOGIE   French jacks called the go-devil a "mogie".

MOGUL BOX STOVE   An early cast iron box stove popular with loggers.

MONDAY LEG or **MONDAY MORNING DISEASE**   A disease of horses accustomed to work and exercise but left tied up in the barn for a couple of days. It made their legs swell. Same as black water.

MONEY MAKER   Same as a scale rule, log scale rule, cheat stick, robber's cane, thief stick, swindle stick.

MONITOR   A windlass used in recovering sunken logs or to tie up a boat. Same as catamaran, pontoon, sinker boat.

—M—

MONKEY   Burnt sugar used by cook to color gravy a rich dark brown.

MONKEY JACKET or **MONKEY SHIRT**   A heavy wool shirt with the tails hemmed up to about hip length. Worn outside, not tucked into pants.

MOOLEY   A cant hook. A tool like a peavey but having a toe ring and lip instead of pike to handle logs on land. Same as cant hook, crooked steel, swing dog, cant dog. Sometimes spelled muley. A mooley cow can't hook because she has no horns.

MOOLEY HOGAN   A person expert with a mooley.

MOONER   A mythical creature in logging woods.

MOOSE BIRDS   Canada jays or garbies. The reincarnated souls of dead lumberjacks. Same as camp robber.

MOOSE CAT   1. A log that goes up onto the deck or landing in the proper manner. When a log goes crooked, it takes time to straighten it to fit with the rest of the pile. 2. Anything unusually large or an unusually good lumberjack.

MORNING GLORIES   Pancakes.

MORNING SHOT   Same as "eye-opener"—a drink of whiskey taken early in the morning after rising.

MORTISE AXE   A poll axe with one cutting edge and the other edge a narrow, chisel-like cutter used for cutting holes through poles and posts.

MOSS   To fill with soft clay, oakum, or mud the crevices between logs.

MOSSBACKS   Name given by loggers to people who come in to settle the cutover lands. Also spelled mothbacks.

MUCKUKS   Bowls or containers made from birch bark.

MUD   1. To fill with soft clay, oakum, or moss the crevices between logs. 2. Coffee or slush.

MUD BOAT   A low sled with wide runners used for hauling equipment from camp to logging operation.

MUD SILL   The bed piece or bottom timber of a dam which is placed across the stream, usually resting on rocks and in mud. Same as bottom sill.

MULAY or MULAY SAW   A stiff, upright saw used generally in water-powered sawmills.

MULE   Alcohol or rotgut whiskey—noted for its kick.

MULE SAW   Same as muley saw.

MULEY MILL   Same as mulay mill.

MULLIGAN   A stew made with meat and vegetables served often in a lumber camp.

MULLIGAN CAR   A railroad car where midday meal was served.

MURPHYS   Potatoes.

MUSHING INTO TOWN   Jacks going to town on a spree on snowshoes.

MUSHING OUT   Movement of lower logs in a pile at a rollway toward the water as the pile was broken out.

MUSH LOG   The outside log of a load of logs.

MUSTARD DRIVE   After the spring breakup and after a "spree", some lumberjacks, especially those from Minnesota, would go to the Dakotas and pull mustard out of the wheat fields. Some would stay on for the wheat harvest.

MUZZLE LOADERS   Old-fashioned bunks into which the lumberjack crawled in from the foot of the bed. Made of flattened poles or rough lumber with a mattress of straw or balsam boughs. Same as shotgun bunks.

–N–

Evening Chores

Lumberjacks worked in the woods until it was too dark to see. After walking back to camp, often one to three miles away, they removed their boots or packs, their wet clothing and red wool underwear, and two or three pair of wet wool socks, which they hung on a balsam holder attached to the big rack over the hot stove. Then they put on Sunday shoes or stag shoes—packs with the tops cut off to make a kind of slipper or brogan.

At one end of the bunkhouse was a long sink with three to ten washpans, according to the size of the crew. For a camp of eighty men, if the bull cook was industrious, there would be about three towels to wipe on, which would not meet sanitary standards today. Water for washing was kept boiling hot by running it through pipes in the big heater to a barrel. The camp furnished soap, generally lye soap made at the camp.

By that time supper would be ready, and what mounds of food the lumberjack could store away!

It was said that steam poured out of the skylight at night like smoke or steam from a factory.

999   One of the jacks' favorite chewing tobacco.

NAIL HEADER   A hand-made blacksmith's tool used to pull nails out of wood.

NAIL KEGS   Spittoons made by sawing in half nail kegs; used in the camp bunkhouse.

NARROW GAUGE   A railroad track no wider than four feet eight and a half inches. The most common was three feet.

NEAR OX   The left ox of a team of oxen. Off ox was the one on the right. Oxen were driven from the left side.

NECK YOKE   A bar with a ring at both ends and in the middle, suspended from the harness collars to hold the tongue of the sleigh in hauling or backing up.

NEEDLE GATE   In a logging dam sluiceway, narrow timbers or poles with two or more squared faces which are placed in contact across the opening of the sluice to prevent the outflow of water. One or more "needles" may be removed without disturbing the remainder.

N.G.   No good, in reference to some cooks. A poor cook.

NICK   to cut a notch to determine the direction a tree would fall. Same as notch, undercut.

NIMROD   A popular plug tobacco.

NIPPER   A member of the steel crew, who, by means of a cross-bar and a block used as a fulcrum, holds the end of the cross-tie against the base of the rail while the spikes are being driven.

NIPPERS or NAIL NIPPERS or NAIL CUTTERS   A sharp cutting pincers used by a blacksmith or farrier to cut the horse shoe nail to the desired length prior to clinching.

NIPS   Pieces of old blankets folded around the feet inside the shoe packs for warmth.

NOONING   Lunch in the woods, so no time was wasted going to and from camp. A hot lunch was brought to the jacks in the woods. Extremely cold weather caused lunches that jacks carried to freeze in a very short time, making very discontented lumberjacks.

NORTHERN PINE   A collective term including both white pine and Norway or red pine.

-N-

NORTON JACK   A large, heavy screw jack used in railroad operations and sometimes in heavy logging operations.

NORTON SCALE   A method of measuring board feet with a scale stick.

NORWAY PINE   Red pine or Pinus resinosa.

NOSE or NOSED   To round off the end of a log in order to make it drag or slip more easily. Same as snipe. A log which has been rounded off at the end for ease in skidding.

NOSE AUGER   An auger with a down-cutting bit edge for boring large holes in wood.

NOSE BAG   1. Tin lunch bucket carried by a river driver, generally strapped to his back. Same as lunch sack, river pig nose bag. 2. A grain bag hung under the nose of a horse when he was fed on the job.

NOSE BAG SHOW   A camp where midday meal was taken to the woods in lunch buckets. Not highly thought of in these latter days.

NO-SEE-UMS   A tiny fly found in the North Woods with a terrific bite. Named by Indians who said, "Could feel um but no see um."

NO TALKING RULE   By order of the cook, therre was no talking permitted at the eating tables. Talking wasted time.

NOTCH or NOTCHING   To cut a notch to determine the direction a tree would fall and to prevent splitting. Same as nick, undercut.

NOTCHER   A man expert in falling trees who notched a tree to make it fall in the desired direction. Incorrectly spelled knotcher.

NOTCH UP   To open the throttle on a steam locomotive.

NUMBER 2   3rd grade lumber.

Prices in the Nineteen Hundreds

RAILWAY TIES—Broad-axe hewed with bark left on, sold to railways at 12 cents for hemlock, tamarack, and cedar.

HEMLOCK LUMBER—Sold at $9.25 per M. (1000) board feet.

BIRCH VENEER LOGS—Gun-barrel straight without a visible defect $13.00 per M. feet, log scale and scaled the smallest diameter of logs.

BEEF QUARTERS—Delivered to northern Wisconsin from Chicago packing houses for 4½ to 5 cents per pound.

OAKEN BUCKETS   Water pails made of pine wood.

OAKUM AND TAR   The standard mixture (loose hemp fibre) for bateau and boat seam calking and chinks in camp buildings.

OAR   A huge sweep to guide a log or lumber raft.

OAR STEM   A pole 30 feet long attached to a grub pin at the stern of a log or lumber raft. The blade of the oar was attached at the water end of the oar stem and thus tahe raft could be guided downriver.

ー○ー

OFF HIS FEED A lumberjack sick in camp.

OFF OX The right ox of a team of oxen. Near ox was the one on the left side. Oxen were driven from the left side.

OFFICE Outhouse, toilet.

OIL BOTTLE Usually an old whiskey bottle filled with kerosene for swabbing a saw gummed with pitch. A notched cork permitted the oil to be sprinkled on the saw blade.

OIL CLOTH Conventional material for a tablecloth in cook shanties. It was easy to clean.

OLD CHOPPINGS Forest areas logged 10–15 years previously and then being logged again.

OLD GROWTH Uncut, virgin tumber.

OLD HEAD A jack who had been around logging camps many years; an old-timer.

OLD MAN The logging superintendent.

OLD PILL A camp doctor.

OLE Short for oleomargarine.

ONE- ARMED PETE A water pump.

ONE-BUNK DRAY A dray with a single bunk used for skidding more than one log at a time.

ONE-HORSE YARDING ROAD A skidding road whereon only one horse is used in the skidding operation.

ON HIS FACE Getting credit with no down payment or security.

ON THE SKIDS From skid road vernacular, slipping, not as good as he used to be.

ON THE STUMP Standing timber.

ON THE TOBOGGAN On the down-grade. A poorly kept camp.

OPEN FACE PIE A custard pie.

OPEN SIDE BLOCK   A block with a hinged side plate permitting a line to be placed in the sheave without threading. Same as a knock-down block.

OTTER SLIDE   A saloon in the basement of a building.

OUTFIT   A logging company.

OUTHOUSE   Privy. Small outdoor building used as a toilet. Camp outhouses were rough shacks with a long pole from which the bark had been removed with a draw knife for a seat, strategically placed parallel to the pit below, not the "last word" in comfort but practical. Many early camps merely had the pole placed parallel to the pit with no shed or roof over it.

OUTLAW   Logging sleigh having bunks longer than fifteen feet, once outlawed by the state because of overloading of teams.

OUT OF THE WOODS   In the clear. Expression used when loggers had finished a part of an operation and were hastening to its conclusion.

OVER RUN   The difference between the log scale of a quantity of timber and the lumber scale of sawed material cut therefrom, usually expressed in percentage.

OVERSHOT WHEEL   Refers to a water sawmill with power being generated by a water wheel turned by water flowing over the wheel, not under it as in an undershot wheel.

OX   In the early pine days, the ox was the chief power for moving logs to landings. They were generally steers, but cows or bulls were sometimes used. The history of transportation of logs in the woods would indicate that oxen came first, then horses, then the steam hauler, and then the tractor and other equipment.

Many camps used horses for hauling and oxen for skidding. Oxen were slower and steadier for skidding, but horses were faster for sleigh hauling.

Occasionally, when the camp was short of meat in the spring of the year, the cook would butcher one of the oxen, but the lumberjacks claimed that the meat was so tough they could not get a fork in the gravy.

—O—

OX BALLS The metal balls placed over the tips of the horns of an ox.

OX BOW Wooden box to hold an ox yoke in place; a frame, bent into the shape of the letter U and embracing the ox's neck as a kind of collar with the upper ends passing through the bar of the yoke.

OX BOW PIN A wooden pin to hold ox bow to yoke.

OX FORK A short-handled fork with two broad lines used in driving hay into the sleigh ruts to keep a sleigh from going downhill too fast.

OX GOAD Same as goad stick.

OX HARNESS In later years, a harness similar to a horse harness was used with oxen in place of the conventional ox yoke.

OX PRONE A frame used to hold ox's foot while it was being shod.

OX SKINNER Same as ox teamster.

OX TEAMSTER A man who drove ox teams in the pine days.

OX TENDER The barn boss. A man who took care of oxen and stable in the logging camp.

OX YOKE A wooden block shaped to fit over the neck of oxen to push against when pulling.

OX YOKE SAFETY PIN Metal pin similar to the common safety pin used to hold the bows of an ox yoke in place.

Pimp Sticks

Lumberjacks smoked pipes, not cigarettes. Bosses would not even hire lumberjacks who smoked cigarettes, which they called pimp sticks. In one Canadian camp, two young men who rolled their own and smoked Bull Durham tobacco asked the mill owner for the makens and the wheelings. He told them, "You are some more them "kido" boys that smoke them paper pipes. You can go the office and get your walking papers."

PACK   Any container for lumberjack gear. Same as turkey, packsack.

PACKING A BALLOON   Carrying one's blankets from camp to camp.

PACKING A CARD   The lumberjack was a member of a union.

PACKING THE RIGGING   To wobblies, a man who was carrying I.W.W. organizing supplies—literature, dues, books, etc.

PACKS   Winter footwear worn by lumberjacks. Made of rubber bottom with leather tops and worn with several pairs of wool socks. Shoe packs. Walking rubbers.

PACKSACK   Any kind of canvas sack in which jacks carried their personal gear. Same as turkey, pack.

PADDLE WHEELERS   Stern-driven river boats used to pull log or lumber booms.

—P—

PAINT A LOG   To indicate with paint the ownership of a log. See paint mark.

PAINT PUN   A later day, automatic paint sprayer used for marking cutting areas, boundaries, etc.

PAINT MARK   Variously colored paints used in the early days of pine logging and river drives to mark logs for ownership. Paint marks were supplanted by end and side marks made with axe or stamping hammer.

PALING   A short board or piece of wood.

PAN   A large flat metal plate curved up in front upon which the front end of logs were placed to make hauling easier and to prevent ends from rutting or digging up the trail. More commonly used when tractors replaced horses.

PAN SKIDDING   Latter day skidding with the use of a large, steel, skidding pan pulled by a caterpillar tractor.

PANNIKAN   A small pan or cup for tea or coffee.

PAN ROADS   Roads built for tractor with pan attached. In recent years tractors were generally used in place of horses and a pan was attached on which one end of a log was placed for skidding logs out of the woods.

PANTS RABBITS   Body vermin. Lice, crumbs.

PARBUCKLE   Line or chain is passed under and back over a log causing the log to roll when the chain or line is pulled. This became a loading method at landings when railroads began to haul logs to the mills.

PARDNER   As in a set of fallers, two men working together as a team at a particular job in the woods. Partners.

PARKA   A winter jacket popular in the woods after 1900.

PASS BOOKS   A form of time slip or due bill issued by some lumbermen to be used only at the company store.

PASS LINE   The line by which a high rigger moves up and down at his work, once the tree is topped.

PASS THE HAT   Taking up a collection for any reason, generally for a jack killed or hurt, or passing the "you know" pan.

PAT HIM ON THE LIP   to thrash or whip a person.

PAUL BUNYAN   The mythical lumberjack. A Hercules among lumberjacks.

PAVING BLOCKS   Wood blocks used for surfacing city streets. They were generally soaked in creosote.

PEA JACKET   The standard overcoat of a lumberjack. A short wool plaid coat. Same as mackinaw, reefer, jumper, frost fooler.

PEAKER   1. A load of logs narrowing sharply toward the top, and thus shaped like an inverted V. 2. The top log of a load.

PEA PICKER   A farmer who worked part time in a sawmill.

PEARL DIVER   A dish washer.

PEA SOUP   A French Canadian lumberjack.

PEAVER   A log driver who rolled the logs from a bank or obstruction into the river during a log drive. Same as sacker.

PEAVEY or PEAVY   Somewhat similar to a cant hook but having the end armed with a strong, sharp spike. For rolling and handling logs in water, whereas a cant hook is used in handling logs on land. Named after Joseph Peavey of Stillwater, Maine, who invented it in 1858. His gravestone in Bangor, Maine, has two crossed peavies carved on it. Also called peewee, crooked steal, Quebec choker.

PEDE   The handcar or velocipede used on the railroad. Also Jemmy John.

PEEL   1. To remove bark as in the hemlock bark camps. 2. To remove the bark from timbers to be used as poles or piling. 3. To peel off bark as in a popple operation for pulpwood.

PEELER   1. One who peels bark from a log. Generally one who peels bark in gathering tan bark. Same as barker or spudder. 2. A veneer log.

—P—

PEELING CHISEL   A chisel to remove bark. A spud, barking spud, or barking iron.

PEELING TIES   Removing the bark on railroad ties.

PEEN HAMMER   A small hammer carried on the harness used to knock the packed snow from the hoofs of the horses. Same as snow knocker, ball hammer.

PEERLESS   One of the older, very "strong", popular and well-liked smoking tobaccos. Could be chewed.

PEEWEE   A peavey.

PEG WOOD   Hardwood used to make rafting pins or grub pins.

PENCIL CEDAR   Cedar stock used in making pencils.

PENCIL PUSHER   The camp clerk.

PENGER, PANGA, or PENGA   A Norwegian word for money or pay.

PENGI   A Swedish word for pay.

PENNY DOG   An assistant foreman.

PERCHES   Jacks' name for Percheron horses.

PERFUME PIE   A pie made from a packaged pie filling.

PER THOUSAND   Refers to logging cost as so much per thousand feet. A unit of measure of timber value.

PERUNA   A patent medicine favored by jacks consisting of beef broth, wine and iron.

PHOENIX LOG HAULER   A treaded type, steam log hauler manufactured at Eau Claire, Wisconsin.

PIANO TRIMMER   Man who operated the trimming saw when boards came from the head saw.

PICK   A heavy pointed iron or steel tool on a wooden handle inserted in an eye between the ends and used to pierce, break up, or free matter such as rock, earth, etc.

PICKAROON   A tool used in pulling small timbers out of the water or in loading ties on cars. Broken axes were sometimes made into pickaroons. Same as hookaroon.

PICK AXE   A heavy tool with hoe blade on one side and axelike blade on the other. Same as mattock.

PICK-HAND SPIKE   Same as pike pole or pike.

PIKE POLE   Same as pike or pike pole.

PICK THE REAR   To roll logs into the water which have lodged or grounded during a river drive. Same as sack the rear.

PICKETS   Fence stock cut at a sawmill.

PICKS   Heavy, hand pikes or pike poles.

PICK UP CREW   Same as "hoist crew". A group of jacks which moved along a logging railroad loading logs at small landings along the tracks.

PIECE CUTTER   A lumberjack who cut logs or pulpwood at so much a stick.

PIECE MAKER   Same as piece cutter.

PIECE STUFF   Lumber cut to specific lengths, often by special order.

PIECE WORK   Same as piece cutter.

PIE FORK, PIE LIFTER, or PIE PEEL   A fork-shaped, long-handled gadget used in lifting hot dishes, especially pies, from the ovens of the cook shanty stove.

PIE IN THE SKY   Wobbly reference to the bourgeois heaven.

PIE JOB   Easy work.

PIE KEEPER   A large wooden box approximately 6'x4'x3' kept in the cook shanty used to store food from flies, especially during summer logging.

PIER DAM   A dam built from the shore slanting downstream to narrow and deepen the channel or to guide logs past an obstruction, or to throw all the water on one side of an island. Same as wing dam.

—P—

PIG IRON Hardwood logs, as they were a lot heavier than pine, spruce, or hemlock.

PIG'S EAR A cheap saloon in a logging town.

PIG'S EYE Original name for St. Paul, Minnesota.

PIG'S FOOT JAMMER HOOK A loading hook shaped like a pig's foot, that is, an iron, clawlike hook used in jamming logs in flatcars. Same as loading hook.

PIG TAIL HOOK An iron hook shaped like a curled pig's tail.

PIKE A supply road to camp. Same as tote road, portage road, pike pole.

PIKE POLE A long pole, twelve to twenty feet long, with a sharp spiral spike and hook on one end, used to handle floating logs. A jam pike.

PILING or **PILES** Long, straight poles or pilings, generally those driven in the ground. Can mean simply a pile of logs, pile of lumber.

PILING RAILROAD A railroad built in trestle form on piling.

PILOT Same as rafting pilot.

PIMP STICKS Cigarettes. Loggers and lumberjacks despised men who smoked cigarettes, and many foremen would not hire them. Their other names for cigarettes are unprintable.

PINCH BAR 1. An iron bar used to pinch logs tightly together. 2. A bar used to move logging railroad cars short distances.

PINCHED Often referred to a crosscut saw gummed up with pitch and stuck in the tree cut.

PIN DOTE Small rotten spots on the ends of logs.

PINE The eastern white pine, which is different from southern yellow pine or western pine. The pine of the Great Lakes area collectively used to include the Norway or red pine.

PINE CHIMNEYS   Tall, (10–30 feet) burned-out stubs of white pine after a forest fire would burn for days.

PINE COUNTRY   An area of the country with heavy growth of white pine. Same as pinery.

PINE GROVES   Contrary to popular belief, the original forest cover was not an unending stretch of pine from the Atlantic Coast west through the lake states; rather, pine grew in groves, which the botanists call "islands of pine". The early white pine logger called these stands or groves.

PINE HOG   A timber baron.

PINE PLATTERS   Wood meat platters made by many wooden ware factories.

PINERY   Same as pine country.

PINERY BOY   Shanty boy or woodsman.

PINERY BOYS   Lumberjacks or shanty boys.

PINERY ROAD   A road that leads to the pine woods.

PINE SLAB BUNKS   bunkhouse bunks made from pieces of pine split but with the bark left on.

PINE TONGS   An extra large skidding tongs opening up to accommodate a pine log five feet in diameter.

PINE TOP   Intoxicating liquor.

PIN MAUL   A heavy, short-handled hammer used to drive wood pins in lumber and log raft construction.

PINUS RESINOSA   Scientific name for Norway or red pine.

PINUS STROBUS   Scientific name for white pine.

—P—

PIN WHACKERS or **PIN WHACKER** 1. men who drove in pins of metal or wood attached to chains or driven over ropes in setting up log booms. 2. Usually a light weight man or boy who worked in one of the many sorting pens or pocket booms at the booming grounds. His job was to stand on each log as it floated into the pen, and with a wooden mallet, drive a staple-like hardwood rafting pin over a length of rope, pinning it to the center of the log. As each log entered the enclosure, it was pinned, roped and drawn up next to its neighbor—thus a raft was made—ready to be towed to the sawmill.

PIN WORM HOLES Small holes in timber and lumber made by the larvae of certain beetles.

PIPE MONKEY A man who connected and repaired pipes on a donkey engine.

PISS WAGON Same as icing sleigh or sprinkler.

PITCH Sticky, resinous material found in coniferous trees. Woodsmen often used it medicinally.

PITCH POCKET A cavity in wood filled with resin.

PITCH SEAM Same as pitch pocket.

PITCH STREAK In coniferous woods, a well-defined accumulation of pitch at one point.

PITCHING IRON A flat iron ladle for stirring pitch and applying same to seams in river boats such as bateaus.

PITMAN A connecting rod between the water wheel and the sash in a water-powered sawmill.

PIT SAW or **PIT SAWING** A two-man, cross cut saw with no raker teeth used to saw boards over a pit with one sawyer standing on an elevated platform and the bottom sawyer standing in a pit. Used mostly in colonial times.

PIT SAWYER The sawyer standing in the pit in a pit sawing operation.

PLANING MILL A mill where lumber is dressed before shipping.

PLANK   To prepare fish (as was done in some of the very early "shanty" camps) by baking it on an untreated, hardwood board, on which the fish was served for supper.

PLANK ROADS   Roads around the sawmill and near towns and cities or between towns. Made from rough-sawed heavy lumber of inferior species.

PLAT BOOK   A detailed map showing ownership and acreage of every parcel of land within a county.

PLAY HELL WITH   To make a bad mess of anything.

PLAYED OUT HORSE   A tired or crippled horse.

PLOWBOY   One of the jacks' favorite chewing and smoking tobaccos.

PLUG   1. Logs caught on an obstruction during a river drive. Same as jam, log jam, side jam. 2. One of the jacks' favorite chewing tobaccos.

PLUG A BOOM   Placing a wooden plug in a hole in a boom stick to hold the boom chain in place.

PLUG AND KNOCK DOWN   A device for fastening boom sticks together in the absence of chains. It consisted of a withe secured by wooden plugs in holes bored in the boom sticks.

PLUG BOOM   A boom of logs connected by ropes passed through holes bored in the ends of the logs. Wooden plugs driven into the holes wedged the ropes that pulled small bunches of logs to the main boom.

PLUG TOBACCO   Chewing tobacco sold in the van. It came in long strands and blocks and was clipped off with a tobacco cutter.

PLYMOUTH SWITCHER   A small switch engine used on short hauls, switching and ballasting.

POCKET BOOM or POCKET   A boom in which logs were held after they had been sorted.

POCKETS   Log pens at a sorting works.

154

POCKETS OF PINE   Scattered stands of pine within a deciduous forest.

POINT CATTLE   Oxen are worked in teams of from 3–5 yoke. The 4th pair is called the "point cattle".

POINTER   Very long-tipped bateau. Generally a Canadian term.

POKE LOGAN   A bay or pocket into which logs may float during a drive. Same as logan.

POLE   The tongue of a wagon or sleigh.

POLE CATS   Railroad tie men who lived far back in the woods away from the main camp.

POLED AND TIED   A simple method of railroad building using corduroy poles placed over a swampy roadbed with the ties being placed on top of the poles in making the corduroy.

POLE LAMP   Same as post light.

POLE ROAD   A road like a railroad track built of ten-inch or larger logs, doweled together at the end and roughly dressed on top to fit iron concave wheels on which logs were carried to the river or the landing on railroad-type cars. These roads were generally built over swamp areas. The logs were greased, particularly at the corners, for ease of hauling. A tram road.

POLE TIE   A tie made from a stick of timber yielding only one tie.

POLERS   Men who direct pulpwood and logs from rafts in the water to the proper location for loading on cars.

POLE SHELF   A shelf 6–7 feet wide constructed along a wall of the bunkhouse a foot or two above the floor. Balsam boughs or marsh hay was placed on this shelf. Jacks in these early camps slept on this shelf with the entire group being covered by a single, long blanket.

POLE TEAM   The team nearest to the load. Same as butt team, wheelers.

POLE TRAIL   A trail across a swamp made by felling Tamarack and black spruce trees (poles) upon which a jack could walk without getting wet.

POLICE GAZETTE   One of the favorite reading materials in the old camps.

POLL AXE   Single-bit axe for chipping and driving wedges and for releasing a cross cut saw from binding when trees were being felled. Used in felling trees in early logging. Same as single-bitted axe.

POND   Any water area into which logs were dumped, as mill pond, sorting pond, log pond.

POND BOSS   Man in charge of a booming operation or a mill pond operation.

POND MAN   One who collected logs in the mill pond and floated them to the gangway.

POND MONKEY   A jack who works at a log pond steering logs to the bull chain which pulls the logs into the sawmill.

PONTOON   A small raft carrying a windlass and grapple, used to recover sunken logs. Same as catamaran, monitor, sinker boat.

POOL DRIVE   The pool was an association of lumber companies that pooled their logs into one drive.

POOR BOX   Box near the door containing free tobacco. Men guilty of misdemeanors were punished by fines paid in tobacco for the other men to use.

POPCORN CREW   A loading crew that made up a poor load of logs on the sleigh, a load which was irregular and might fall off the sleigh bunks while being transported.

POPPLE   The aspen tree. Natives and jacks always called the aspen popple.

PORCUPINE BOARDING HOUSE   A tree with an excessive hollow in it.

PORTAGE ROAD   A supply road to the camp. Same as tote road, pike.

━P━

PORTAGE TRAIL A trail used to carry in supplies in a packsack.

PORTAL TO PORTAL From bunkhouse to woods before and after daylight.

PORTER Name of an early steam locomotive used in woods operations.

POST LIGHT A large, kerosene lantern mounted on the top of a pole in the camp yard area giving some light for walking at night.

POT A donkey engine.

POT-BELLIED STOVE A rather squat, round stove used to heat bunkhouses.

POTATO HISTERS What the old-time lumberjacks called farmers from the southern part of the state who worked in the woods in the wintertime.

POTTER A round stick, 3 inches to 4 inches in diamter and 2½ feet to 3 feet in length with an iron clasp in the center which is fastened to a short piece of chain with a hook at the free end. The hooks are fastened to the decking chains in loading and the potters hold the load in a vertical position.

POTHOLE A small woods pond, sometimes a small lake.

POT WALLOPER A cook.

POUND THE TIES Quitting, a worker often walked the track to town or to another camp.

POWDER Black blasting powder or dynamite.

POWDER MONKEY Man in charge of blasting operations, usually when breaking a log jam in a stream.

PREGNANT WOMEN Dried apples. Sometimes called Adam's fruit.

PRIME LOG In the export market, one that is free from defects.

PRITCHEL A punch which has been drawn to a small rectangular point used to punch nail holes through horse shoes.

PRIZE LOADS   A very large load of logs on a sleigh usually done for "picture-taking".

PRIZE LOGS   1. Logs which came to the sorting jack without marks denoting ownership. Same as scrabble logs. 2. Something the jacks loved dearly or prized highly—wife or sweetheart.

PROD   A stick with a sharp point on the end used by ox teamsters in early pine days. A goad stick.

PROTESTANT CROSS   A mattock.

PRUNE BURNER   A blacksmith.

PULASKI   A tool used in fire fighting. A combination tool equivalent to an axe and hoe. A mattock. Sometimes called a hodag.

PULL BOATS   Steam tugs used to push or pull log or lumber rafts.

PULLED IN   Logs taken from a river at a rafting chance.

PULLING THE OAR   The man at an oar (sweep) on a lumber raft would raise the stem (handle) of the oar so as to push the blade of the oar as deep as possible into the water, and would then take five steps as he pushed the oar. He would then drop the oar by letting go of the stem and swing back to his original position. Thus, this was pulling the oar; on returning, it was pushing the oar.

PULL THE BRIAR   To use a cross cut saw.

PULL THE PIN   1. Slang used by rail crews referring to pulling the pin in a link and pin coupling. See link and pin. 2. Also, to suddenly walk off the job or quit.

PULLEY   Same as block.

PULL UP   A place on the bank of a river, lake or pond where logs that had been out and floated down from the woods were pulled out of the water to be milled into lumber or loaded on a train for shipment to a sawmill.

PULL SAW TRIMMER   A circle saw, mounted on a movable frame, which could be manuevered by hand as a trimming saw.

–P–

PULP   Same as pulpwood.

PULP HOOK   A short, one-hand tool for handling short bolts or pulpwood. Bolt hook. Birch hook. Wood hook.

PULP RACK   A wooden box-like frame built on sleigh bunks used for hauling hemlock bark.

PULP RULE   A scale stick used in determining a volume of pulp wood.

PULP STICKS   Individual pieces of pulpwood.

PULPWOOD   Wood cut for paper mills, usually in 100-inch lengths for truck hauling. Before the use of trucks, it was cut 144 inches long.

PULPWOOD CHECKER   A scaler who checked and recorded the volume of pulpwood stored at a landing.

PULPWOOD STANDS   Standing trees to be harvested for pulpwood.

PUMPER   A handcar.

PUMPKIN PINE or PUNKIN PINE   Same as cork pine.

PUNCH HAMMER   A short-handled, long-nosed hammer used by blacksmith to enlarge holes in hot iron.

PUNCHEON FLOOR   In the very early camps the log floors were hewed smooth with a straight adze.

PUNCHEONS   See puncheon floor.

PUNCHINGS   Jacks' term for puncheon floors.

PUNG   A crude oblong box on poles that served both as the shafts and the sled runners. Originally used by Indians and pioneers in hauling.

PUNK   1. Any young man, but specifically a whistle punk. 2. A bad spot in the wood cause by an injury to the tree when it was young. Same as blind punk. 3. Bread. 4. Wood so decayed as to be dry, crumbly, and useful for tinder. 5. A material sometimes used to light fuses on dynamite charges.

PUNK BOY   A chore boy.

PUP HOOK   A small hook at the end of a loading line or chain.

PUSH   The camp foreman, woods boss. Same as big push.

PUSH CART   A four-wheel, flat decked car used on railroad track repair.

PUSHER   Same as push.

PUSHING THE OAR—PULLING THE OAR   The tiller motion of the steersman on a log or lumber raft.

PUT ANOTHER HOLE IN THE BARREL   After the passing of the old camboose with the smoke hole in the roof, round holes were cut into the bunkhouse roof and barrels, with the bottoms still in and holes for ventilation, were placed upside down over these holes. If the bunkhouse got too hot or too ripe from the odors of drying wet socks and other clothes, one of the jacks would call out, "Put another hole in the barrel," meaning that more ventilation was needed.

PUT IN   To deliver logs at the landing.

PUT IN LOGS   Cut logs.

PUT ON THE FEED BAG.   To eat.

PUT THE CALKS TO HIM or **PUT THE BOOT TO HIM**   In a barroom brawl, to stomp on a man's face.

PYRAMID   A heavy iron cone used in blacksmith shop in shaping rounds of all kinds. Same as cone, mandril.

–Q–

Sleeping in the Bunkhouse

The very early camps had two tiers of bunks built on both sides of the men's camp. They were called muzzle-loading bunks because the jacks crawled in from the end—two men to a bunk with a snorting pole between them. The bottom of the bunk was sometimes made of rough lumber but generally of poles. The "mattresses" were of balsam boughs or straw.

Jacks preferred the top bunk because sleeping below meant having straw falling down in one's face. The logging company usually furnished blankets. Later, ticks filled with straw were used. Mattresses harbored bedbugs, but straw could be taken out and burned.

On the outer end of the bunks was the deacon seat. It was made of lumber or the smooth side of half a log. It provided the only "chair" in the bunkhouse.

Generally the only argument at night was whether it was too hot or too cold. One sleeper would get up and open the skylight to cool the room, and someone else would get up and close it. Often the bull cook had to settle the argument, and he generally kept it too hot. Skylight ventilation was preferred because it lessened the odor of a hundred socks drying around the big stove in the middle of the bunkhouse.

When the logging in an area was finished, cooking equipment and stoves were removed and the log building was left to rot or burn in a forest fire.

QUARTER SECTION   One-fourth of a land section, 160 acres.

QUEBEC CHOKER   A peavey.

QUENCHING TANK   The blacksmith's cooling tank of water.

QUICK WATER   The current in a stream fast enough to make white foam. Same as white water, rapids water.

QUININE JIMMY   A camp doctor.

–R–

River Drive

Throughout the drive, the drivers were often wet the entire time. Sometimes they rode the logs; sometimes they walked along the bank on a path called the gig trail. Trout fishermen still use these old trails. The men slept where they could, often on the ground under the stars, or on a bed of balsam boughs. Some lumberjacks were superstitious and would not change wet clothes but let them dry on their bodies for fear they would get pneumonia. Following the drive was a bateau or a wanigan, carrying food and supplies. The bateau was a flat-bottomed boat tapered at each end. The wanigan was the cook's raft and was generally used on large rivers.

A long drive could last several weeks, part of which was without contact with the outside world. Generally, however, at a few points tote roads ran to the river where fresh supplies could be brought in. Food was carried in a "river pig's nose bag" strapped to the back of the river driver.

The river drive was probably the most dangerous of any occupation in the history of the state. Many river drivers (or "river rats" or "river pigs") were drowned. Many more had legs or bodies broken. A driver who met death was buried on the bank, and often no stone marked his grave. His shoes were removed and hung on a tree near his grave. Sometimes the tree would be marked. He was simply forgotten. Injury or death was considered a part of the job. Regardless of danger, the logs had to move.

RABBLE   The fire shovel used at the forge in a blacksmith shop.

RACK   A sled body for hauling short bolts such as pulpwood.

RACK A LOAD   Teamster's expression for breaking loose frozen runners by making the team go right and left.

RACK A SLEIGH   Same as rack a load.

RACK FLATS   Railroad cars used for hauling pulpwood having vertical stakes at each end of the cars.

RACK THE POLE   Moving the tongue back and forth by geeing team to the right and hawing them to the left in order to break the runners loose. The cross chains broke loose the rear runners by this process.

RAFT   A unit or raft of logs on a river or lake.

RAFT FLAG   Lumber and log rafts flew an ownership flag at all times.

RAFTED OUT LOGS   No more logs. River drive completed.

RAFTER DAM   A dam in which long timbers are set on the upstream side at an angle of from twenty to forty degrees to the water surface. The pressure of the water against the timbers held the dam solidly against the stream bed. Same as self-loading dam, slant dam.

RAFTERS   1. River pigs. Generally those who rafted lumber rather than logs. 2. Raft boats used to tow logs and lumber.

—R—

RAFTING ANCHOR   Any anchor used in a log or lumber rafting operation.

RAFTING AUGER   Very long auger with an offset handle for drilling holes in raft logs.

RAFTING CHANGE   Space along a river, usually with high bank and deep water, where logs were delivered and made into rafts. Rafting grounds.

RAFTING CHUTE   The floor of the rafting table which could be tilted to allow lumber cribs to slip into the river to be made into lumber rafts.

RAFTING DOG   Large, heavy-eyed spike driven into outer logs in making secure bindings in a raft of logs.

RAFTING GROUNDS   River area where log rafts were assembled. Rafting chance.

RAFTING IN   Constructing lumber or log rafts.

RAFTING PICKET   A lane lined with catwalks leading out of a sorting picket in a log pond. Logs of the same ownership were steered into rafting pickets to be assembled into log rafts.

RAFTING PILOT   Steersman on a lumber or log raft.

RAFTING PIN   A hardwood wedge driven into the logs of a log raft to which ropes were attached to hold log together.

RAFTING RIGGING   Rope used for many purposes on a drive.

RAFTING ROPE   Heavy, Manila rope used to bind logs together in a log raft.

RAFTING SHEDS   Sheds at a sawmill located near the river used to construct the lumber rafts for shipment downriver.

RAFTING SLED   Sleds on which lumber raft cribs were made.

RAFTING TABLE   Platforms in rafting sheds upon which lumber cribs were constructed. The tables could be tilted to allow cribs to be pushed into the river in making up lumber rafts.

RAFTING WORKS   Booming grounds. A place where logs were held.

RAFT OAR   An oar for a log or lumber raft approximately 45 feet long with a stem being approximately 30 feet in length with an attached blade made from a 16 foot plank, 14 inches wide and 3 inches thick. This huge oar was used in tiller-fashion to guide the rafts downriver.

RAFT SHACKLES   Two spikes connected by a short chain. Same as rafting dog.

RAILHEAD   Farthest point in the woods to which the railroad had been laid.

RAIL KINKER   A railroad brakeman.

RAILROAD CAMP   A logging camp in which most of the buildings were hauled from the cutting area to another by railroad flat cars.

RAILS   Split oak or cedar used for rail fences.

RAIL TONGS   A very heavy two-man tongs used in moving rails in track-laying and repair.

RAISE HEAD   Closing the sluice gates so that water will rise behind a dam.

RAKE   A deflector placed up-stream from bridge cribs as protection against ice and high water.

RAKER GAUGE   A tool used with a hand file to insure that the raker teeth on a crosscut saw are filed to the right length to work correctly with the cutting teeth.

RAKERS   Clearing teeth in a crosscut saw. The teeth that removed the sawdust.

RAM PASTURE   1. The bunkhouse of a logging camp. 2. A hotel room generally larger than the barroom of a saloon in which lumberjacks slept on cots or on the floor.

RAM PIKE   A standing dead tree from which the limbs and top have been broken or burned off. Same as snag.

166

R

RAN CAMP   Same as bossed camp.

RANK   To haul and pile regularly, such as to rank bark or cordwood.

RANKING BAR   Two poles held in position by rungs upon which bark or wood is carried by two men. Generally used to carry tanbark. Same as hand barrow.

RANKING JUMPER   A wood shod sled upon which tanbark was hauled.

RAPID LOADER   Similar to the Barnhart and the Model C American steam-powered loaders.

RAPIDS PIECE   One, two, or three cribs of lumber. A unit of a raft of lumber that was sent through the rapids at one time.

RAPIDS PILOT   A log drive specialist who guided log and lumber rafts through treacherous rapids.

RAPID WATER   The current in a stream fast enough to make white foam. Same as white water, quick water.

RASP   A heavy, course file used by blacksmiths to cut the hoof wall when shoeing a horse. Also used by the wood butcher to cut wood edges. Sometimes called: "Double-end Horse Rasp".

RATCHET DRILL   One of the few mechanical tools in a camp used for drilling holes in timbers for the insertion of wooden pegs.

RATTLINGS   Sawmill debris.

RAVE   A piece of iron or wood which secured the beam to the runners of a logging sleigh.

RAWHIDE   To hurry. Same as bull 'em through.

RAYMOND JAMMER   A steam log loader working from a railroad track. Similar to the Cody and the McGiffert jammers.

RAYMOND LOADER   A jammer on sleighs with runners turned up on both ends to enable it to travel in both directions. Used to load logs on sleighs.

REACH or **REACH POLE**   A steel shod wooden pole for fastening sleighs together when hauled by steam haulers or tractors.

REAMER or **REAM AWL**   A large spiral bit with sharp corners used to cut ice out of water holes in a sprinkler.

REAR   The upstream end of a drive; the logs either in the water or on the bank.

REAR CREW   The sacking crew following the tail end of a log drive. Those men in a driving crew who cleaned up the river and brought all hung-up logs back into the mainstream.

REARING   Same as sacking the rear.

RECEIVING BOOM   A boom used to hold floating logs in storage at the sawmill. Same as storage boom, holding boom.

RED CARD   A membership card in I.W.W., International Workers of the World.

RED EYE   Whiskey or booze.

RED HORSE   Salted beef; also corned beef.

REDLEAD   Ketchup.

RED PINE   Same as Norway pine.

RED ROT   A heart rot fungus in pine tree trunks causing wood to become dry and flaky.

REEFER   A jacket or short coat made of thick cloth. Same as mackinaw, reefer (in Michigan).

REEFING HER   Pushing a boat with a pole.

REFLECTOR OVEN   A box-like copper or sheet metal oven used to catch the heat and bake bread or biscuits before the open fire. Used by timber cruisers and land lookers.

REFUSE   That portion of a tree that cannot be removed profitably from the forest or utilized profitably at the manufacturing plant.

–R–

REFUSE BURNER Large, conical tower in which refuse such as waste ends, sawdust, chips, and bark were burned at a sawmill.

RELOG To pick up small logs left over from the logging operation. Same as salvage logging.

RENDERING Playing out a mooring line from a log or lumber raft when stopping the raft before nightfall.

RESAW A type of mill saw used for cutting lumber into smaller pieces.

RESIN or ROSIN Pitch. A secretion of the evergreen tree. Secretion of balsam fir generally called rosin.

REST POWDER Snuff.

RICE BITS An imaginary tool used to fool greenhorns in a logging camp.

RICE POWDER A face powder used by prostitutes.

RICK 1. A pile of wood cut to stove length, usually sixteen inches, piled four feet high and eight feet long, three ricks to a cord. In twelve-inch length, four ricks to a cord. 2. A pile of cordwood stave bolts, or other material split from short logs; a cord eight feet long, four feet high, and of a width equal to the length of one stick.

RICOCHETS An imaginary tool used by jacks to fool camp greenhorns.

RIDE The side of a log upon which it rests when being dragged.

RIDE A LOG To stand on a floating log.

RIDE POLLY To draw half-pay as the result of an accident.

RIDER A boom stick with an extra hole bored in it about twelve feet from the end of the stick. At the bow of the boom, the head stick was fastened to this hole by chains. At the stern, the tail stick was thus hooked on.

RIDE THE DONKEY SLEIGH Quit the woods job and take a ride back to town with the tote teamster.

RIDE THE SAW   Refers to a heavy and slower stroke on a crosscut saw.

RIDING   The pressure put on a crosscut saw in cutting down a tree. A sharp saw did not need much pressure.

RIDING SHANKS MARES   Walking.

RIDING THE SAW   1. Bearing down too hard on a crosscut saw. 2. Pulling too hard to one side in using a crosscut saw, thus making a curved cut.

RIFFLES   Term used sometimes for rapids. Generally, however, it referred to water not as swift as rapids or white water.

RIGGIN   1. The doubletrees, singletrees, swivel hook with hand ring, and chain or tongs for skidding logs. 2. The cables, blocks, and hooks used in skidding logs by steam power.

RIGGING BLOCK   Steel pulley block used in logging.

RIGGING GANG   Crew piling logs at a landing.

RIGGING SLINGER   Man of a yarding crew who attaches the chokers to the main yarding line.

RIGGIN SWINGER   A man who swung the rigging into place when skidding logs.

RING   A section of tanbark, usually four feet long.

RING BOOM   Small booms used to gather logs or pulpwood that washed over the booms.

RING DOG   Any dog attached to a fairly large ring through its eye through which lines were run. Could also be used to roll logs.

RING ROT   Decay in a log, which followed the annual ring rather closely.

RIP SAW   A hand saw so designed and filed as to cut a longitudinal kerf in logs or lumber.

—R—

RIP TRACK   An "end-of-the-line" track of a logging railroad, usually in the logging railroad headquarters town and used as an area where worn out box cars, flat cars, and other railroad rolling stock was dismantled.

RISE   1. The difference in diameter, or taper, between two points in a log. 2. The crest or high water of a freshet.

RIVE   To split off shingles with a froe.

RIVER BOSS   The foreman in charge of a log drive.

RIVER CONVENTIONS   A series of conventions held by sawmill owners shortly after the Civil War in efforts to have more publicly-financed river improvements on the Mississippi River from New Orleans to St. Paul, Minnesota.

RIVER DRIVE   Taking logs to a mill by floating them down a river. This was the most dangerous part of early logging, and many lumberjacks lost their lives on these drives.

RIVER DRIVER   One who worked on a log drive. Same as river hog, river pig, catty man, river rat, river jack.

RIVER HOG   A lumberjack who worked on a river drive. Same as river pig, catty man, river rat, river driver, river jack.

RIVER JACK   Man who worked on a log drive. River pig, river hog, river rat, river driver, a catty man.

RIVERMEN   Same as rafters.

RIVER PIG   One who drives logs down a river. The lumberjack was a special breed of man but the river man or river pig was very special. It was said good river men were born, not made. Same as catty man, river hog, river rat, river driver, river jack.

RIVER PIG NOSE BAG   A tin lunch bucket carried by a river driver, generally strapped to his back. Same as nose bag, lunch sack.

RIVER PIRATES   Unscrupulous men living along the river banks stole logs at night as timber was being floated to the sawmills downriver.

RIVER POLICE   Local law enforcement officers assigned to police river banks in an effort to stop log thievery at night.

RIVER RAT   A lumberjack who worked on a log drive. Same as river hog, river pig, catty man, river driver.

ROAD BREAKER   Same as swamper.

ROAD GANG   Those members of the crew of a logging camp who cut out logging roads and kept them in repair.

ROAD LOCOMOTIVE   In early days of logging any road engine, a steam hauler.

ROAD MONKEY   A man who kept the sleigh road or truck roads in good shape. Same as blue jay, hay man on the hill.

ROADS WENT OUT   the iced hauling roads melted by a spring thaw.

ROBBERSARY   A sawmill commissary.

ROBBER'S CANE   A long hardwood ruler used in estimating the board feet in logs or standing timber. Same as log scale rule, scale rule, cheat stick, money maker, thief stick, swindle stick.

ROBBER'S STICK   Same as robber's cane.

ROCKER   1. A squared timber with a central swivel to enable front runners of sleigh to turn. 2. The bed of a logging sleigh.

ROCKY MOUNTAIN HUCKLEBERRIES   Prunes.

RODDIS LINE   Main line of the Roddis Lumber Co. in Wisconsin.

RODDIS MUTTON   Venison served in some of the early Roddis Co. camps in Wisconsin.

ROD ENGINE   Mainline logging locomotive with piston rods directly connected to the driving wheels. The opposite of geared types often used on spur lines.

ROD IRON   Pieces of round, square, hexagon or octagon iron of varying dimensions used by a camp blacksmith.

—R—

ROD LOCOMOTIVE   A standard steam engine. Same as rod engine with piston rods. Not geared.

ROLL   1. Bedding. 2. The round piece of wood held in place by a gudgeon pin to hold the sleigh runners in place. Same as roller, upright roller.

ROLL BARK   Hemlock tan bark that has not been carefully dried and hence is of inferior quality.

ROLL DAM   A dam built without gates. The logs floating over the dam caused a roll after which the dam took its name.

ROLLED   Robbed while drunk. Same as frisked.

ROLLER   The cross bar of a logging sled into which a tongue is set. Same as upright roller roll. A cant hook man.

ROLL IN   Go to bed. Hit the sack.

ROLLING   Fleecing or stealing a jack's stake.

ROLLING DAM   Logs implanted in the bank, thereby narrowing up the stream to deepen the water to get logs through a rough spot in the river.

ROLLING STOCK   Doughnuts.

ROLLOUT, ROLL-OUT, or **R-O-L-L-O-U-T**   Roll out, tumble out, any way to get out. The cook's predawn cry to the snoring lumberjacks to get up and get to work. The next call was "Come and get it". Same as "daylight in the swamp". It was claimed that the cook's big voice was developed by hollering in a rain barrel, just as some great singers do before going on stage. These calls caused the sleeper "to tremble and start from the land of dreams to the land of pork and beans".

ROLL THE BOOM   To roll a boom of logs along the shore of a lake, held there by wind, by use of a cable operated by a steamboat or kedge. The cable was attached to the outer side of the boom, hauled up, then attached again, thus propelling the boom, by revolving it against the shore when it would be impossible to tow it.

ROLL THE GUFF   To converse; to talk.

ROLL UP YOUR STUFF AND HIT THE ROAD   "You're fired!"

ROLLWAY   1. Any single tier of decked logs. 2. Logs piled along a river to be rolled into the water. 3. A natural or prepared slope for rolling logs into a stream.

ROLLWAY MAN   A logger who worked on the rollway. In the early days of animal skidding, the rollway man pulled dogs and hooks at the landing.

ROOSTER   1. A bar used to couple two logging sleighs. Same as dray hook, goose neck, bumper pole, slipper. 2. A name given to all river raftsmen.

ROOSTER COMB   A small device used in the hook end of a cant hook to send up logs when loading with a single chain.

ROOT CELLAR   A cavelike depression, usually dug into the side of a small hill or knoll near the camp, shored up and roofed over with log rafters and earth in which perishable foodstuffs such as onions, potatoes, etc. were kept from freezing in the winter.

ROOT HOOK   A very thick, heavy hook varying in shape used to pull tree roots from the camp roads.

ROOT HOUSE   Same as root cellar.

ROPE FED CARRIAGE   Very early sawmills where ropes were used to pull logs against the headsaw.

ROSIN   A sticky secretion of evergreen, generally that of the balsam. Pitch.

ROSS   To hew the end of a log which slid on the ground when logs were to be hauled on a travois so that it would slide easily. To trim a log by removing the bark on one side to make it easier to skid. Same as butting timber.

ROSS A LOG   Debark a log. Same as ross.

‒R‒

ROSSER   1. One who barked and smoothed the ride of a log in order that it might slide more easily. Same as log fixer, slipper. 2. A huge plane used to smooth off hew timbers.

ROSSING   Same as barking.

ROTARY SAW   Same as circular saw or circle saw.

ROT GUT   Cheap whiskey.

ROTTEN KNOT   A knot which is not as hard as the surrounding wood.

ROUGH AND TUMBLE LANDING   A landing where no attempt was made to place the logs in neat piles.

ROUGH LOCK   Chain around the runner of a sleigh which acted as a brake.

ROUGH LUMBER   Lumber not processed through a planing mill.

ROUND FORTY   About forty acres of land. To cut a round forty meant that the cutting crossed the line into the next forty. The practice was to cut to the line plus as far as a man could throw his axe. Same as "he logs on section 37". Some loggers would purchase from the government or state one "forty" in the middle of a large stand of timber and, in logging it, continue to cut the pine timber to the north, to the east, to the south, and to the west, hence the "round forty".

ROUND HEAD   A Scandinavian lumberjack.

ROUND HOOK   A half-round hook through which a chain would slide like a slip noose in a rope.

ROUND KNOT   A knot that is oval or circular in form.

ROUND SAW   Same as circle saw.

ROUND SQUARE or **ROUND SQUARES**   An imaginary tool used by jacks to fool camp greenhorns.

ROUND STUFF   Slang for logs.

ROUND TIES   Round railroad ties resembling pulp stick and not squared off. Were used only on woods spur railroad lines and never on logging companies' mainlines.

ROUND TURN   1. The loop at the end of a sleigh road where the driver could turn around without backing. 2. A space at the head of a logging-sled road, in which the sled may be turned around without unhitching the team.

ROUTE   Total length of stay at any given camp; a long route meant a long stretch of employment on one job.

ROVING FORTY   A small acreage of land owned by a logger who, however, cut timber all over the country with no regard for ownership.

RUBBING SAND   Occurred when a lumber or log raft scraped bottom in transit during a river drive.

RUN or RUN ON A HILL   A sleighload of logs that pushed the horse more than they could hold back with the harness due to improper sanding of a hill. Sometimes horses and drivers were killed on a run.

RUN A LINE   To make a boundary survey, establish a section or quarter-section line.

RUN AROUND   That portion of the stream where the river had changed its course, often because of a log jam.

RUN DAM   Tiers of hewn logs built into the bank at the curve of a driving river. The logs hit it and ran past instead of jamming.

RUNNER   A messenger sent downstream to notify the drivers that a sufficient head of water had accumulated at a dam to drive logs downstream.

RUNNER BLANKS   Curved hardwood pieces often called "Wishbones" from which the runners for go-devils were made.

RUNNER CHAIN   A chain bound loosely around the forward end of the runners of a logging sled, which acted as a brake.

=R=

RUNNER DOG   A curved iron attached to a runner of the hind sled of a logging sleigh. It held the loaded sled on steep hills by being forced into the bed of the road by any backward movement.

RUNNERS   Longitudinal pieces faced with iron (the shoe) upon which sleighs move.

RUNNING A SKID BANK   Not reporting all logs skidded in one day while keeping a number to be added to a poor skidding day.

RUNNING FOOT   Measured in a straight line, as linear foot. Twelve inches or one-third of a yard, as opposed to a board foot.

RUNNING LOGS   Logs drifting downstream in a river drive.

RUNNING STEEL   A problem in rough, hilly country where logging railroads hauled in one direction. Heavy loads moving downhill had a tendency to work the rails down with them and when the steel became tight it would kink to cause derailment.

RUNNING THE COMPASS   Timber cruisers were said to "run the compass" meaning that they would walk many miles merely by taking compass bearings as they walked.

RUNS   Total number of logs driven downriver in a season.

RUN SLED   Any sled or sleigh used to haul logs from the cutting area to the landing.

RUNWAY   A long skid road from woods to skidway. Same as drag road, dray road, gutter road, travois road.

RUSSELL CAR   A short railroad car used to haul logs. Devised to handle one length of logs over two axles and four wheels with no brakes.

RUSTLE WOOD   To bring the wood for the cooking or heating stove inside the logging camp.

RUT CUTTER   Same as rutter or rutter rig.

RUTTER   A form of plow for cutting ruts in an iced logging road for the runners of a sleigh. It was often combined with a snowplow.

The roads were sprinkled with water from the water tank and frozen to make ice roads. Same as swamp angel, groover, gouger.

RUTTER BLADES   Iron blades attached to the snow plow for cutting ruts which were filled with water to make the iced roads over which big loads could be hauled.

RUTTER RIG   Same as rutter or rut cutter.

Lumberjack Hierarchy

A lumberjack generally started his woods work as a road monkey or a swamper. The next winter he might become a conman, sprinkling the roads to make the ice roads for the sleighs, or an axe man or sawyer, falling the big pine. To tend skidways as a cant hook man was a step up the ladder. When a lumberjack qualified for loading by demonstrating efficient use of a cant hook, he was considered in an upper bracket, one of the aristocrats of the lumberjack fraternity. Very often, however, axe men and sawyers did nothing but fall trees. Teamsters often did no other work than skidding or sleigh hauling. They had to like oxen or horses and be able to handle them carefully or their animals would be hurt, and good logging horses were worth lots of money. In more recent logging, teamsters farmed in the summer and would take their own team to the woods in the winter. A team was considered worth about as much as a man in the woods.

One time a college professor asked a lumberjack who was considered one of the best in the North Woods, how he had become such an expert axe man. Looking the professor squarely in the eyes, he replied, "Boss, I guess I just always knowed how."

Some lumberjacks boasted to their dying day that they had been born with an axe in their hand.

SACKER A log driver who rolled the logs from the bank or obstruction into the main current with a peavey during a log drive after the main drive had gone down the river. Same as peaver.

SACKING 1. To roll into the water the logs which have been thrown up on the shore during a log drive. 2. To roll or haul by manpower any logs from shallow to deep water.

SACKING BOAT Boats used by the sacking crew in getting all stranded logs back into the mainstream.

SACKING IN Same as sacking the rear.

SACK RAFTS Loose logs floating in a river.

SACK THE REAR To follow a river drive and roll into the water logs that have lodged or grounded. Same as pick the rear.

SACK THE SLIDE To return to a slide or chute logs that have jumped out.

SACKS Same as log booms.

SADDLE BAG A float of lumber that has become doubled around a tree, a sand bar, or island when going down the river, or balanced sideways on a rock or bar in a river.

SADDLE THE SKIDS To hew cut one side of the skids so that the logs would pass over without slipping sideways.

SAFETY FIRST A camp welfare man.

—S—

SAG Let loose.

SAGINAW Retarding the larger or butt end of a log in loading it up on a car. Opposite of Saint Croix.

SAGS What the loggers in the Upper Peninsula of Michigan called the lumberjacks who first came in from the famous valley of the Saginaw River in lower Michigan.

SAILER A hanging branch of a tree. Same as widow maker.

SAINT CROIX Retarding the small end of a log in loading, generally by using a cant hook on the underside. Opposite of Saginaw.

SALARATUS Baking Soda

SALT LICK A spot where salt was put on the ground or on an old stump for oxen to lick during the summer months when they were "turned loose" to forage for themselves until fall when logging operations began again.

SALUSIVER A match. Same as lucifer.

SALVAGE LOGGING Picking up small logs left after the regular logging operation. Same as relog.

SALVE Butter or oleo. Same as lard.

SAMPSON An appliance for loosening or starting logs by horsepower. It usually consisted of a strong, heavy timber and a chain terminating in a heavy swamp hook. The timber was placed upright beside the piece to be moved, the chain fastened around it, and the hook inserted low down on the opposite side. A team hitched to the upper end of the upright timber then applied leverage.

SAND Sugar.

SAND BIN A box or barrel containing dry sand located near railroad coal bins and used by locomotives for traction.

SAND FLEAS Fleas bothersome to jacks in many summer logging operations.

SAND HILL   Any hill that had to be sanded to hold back a sleigh and prevent sluicing.

SAP IS RUNNING   Signified a tree in springtime, when it grows rapidly. Bark is then easy to peel and remove, as from pulp sticks.

SAP IS STOPPED   A tree in winter dormancy. Bark is tight and hard to remove.

SAPLING   A young tree with trunk diameter no more than four inches. Not merchantable.

SAP PINE   An inferior pine—heavy, knotty, and pitchy compared with the white pine monarchs. A term used mainly in Canada.

SAP STAIN   Discoloration of the sapwood.

SASH SAW   A water-powered saw attached to a heavy sash which in turn is run by a pitman attached to the water wheel. Similar to frame saw.

SATCHEL STICK   A stick carried on the shoulder and used by a lumberjack to support his turkey.

SAWBILL   Daily production demand for a sawmill.

SAW BOSS   A man in charge of a crew of sawyers and swampers in the woods.

SAW BOX   A box, six feet by eight feet, in which new crosscut saws were delivered. Saws were often stored in such boxes and transported from camp to camp.

SAW BUCK   A short log with four legs extending through the log to form an X. Used for sawing logs into stove wood. Same as saw horse.

SAW CREWS   In the pine days, the fallers.

SAW GUIDES   That portion of a band saw apparatus which keeps the saw moving in a straight line.

SAWDUST   The small fragments of wood made by the cutting of a saw. Waste wood.

⸺S⸺

SAWDUST BOX or **SAND BOX**   The bunkhouse spitoons.

SAWDUST HOOK   A short handled rake (with teeth) used in a sawmill to clear sawdust formed around machinery in the sawing operation.

SAWDUST CENTERS   Name given to the larger sawmill cities in the lake states.

SAWDUST CITY   Eau Claire, Wisconsin

SAWDUST EATER   One who worked in a sawmill.

SAWDUST WITH LEGS   Body lice.

SAW-FILING HORSE   A four–six foot beam supported by four legs. Clamps held a cross cut saw to the beam for sharpening.

SAW FITTER or **SAW FILER**   Man who sharpened all the camp saws. Same as filer.

SAW-FITTING   A term used to describe saw filing and general care of a crosscut saw.

SAW HORSE   A short log with four legs extending throught the log to form an X. Used by the bull cook for sawing logs into stove wood. Same as saw buck.

SAWING FAT   Cutting lumber unnecessarily thick, wastefully.

SAWING ON THE LANDINGS   Bucking full-length logs at a landing or log dump into desired lengths.

SAW KERF   The width of cut made by a saw. Same as kerf.

SAW LENGTHS   One-man cross cut saws were made in the following lengths: 1. Usually made in 4 foot lengths. 2. Some were made at 3½ foot lengths. 3. A few were made in 4½–5 foot lengths.

SAW LENGTHS   Two-man, cross cut saws were made in the following lengths: 1. 6–7 feet long: This was the usual length and mostly 6 feet. 2. Very few were made at ten feet in length. 3. Seldom were made in 8 and 9 foot lengths.

SAW LOG   As different from pulp sticks, any log from which lumber could be cut.

SAWMILL   A plant at which logs were sawed into lumber.

SAW OIL   A coal derivative, usually kerosene. sprinkled on a saw edge to cut the accumulation of pitch or other resinous substances which caused a saw to pinch or stick.

SAW OIL BOTTLES   Same as oil bottles.

SAW PULLEY   The pulley used on a shingle mill.

SAW SET   Every tooth of a crosscut saw must be bent over just a tiny bit to give clearance for the blade to cut. This "setting" is done with a light setting hammer or with a plier-like device called a "saw set."

SAW STAND   A sapling 3–4 inches in diameter cut to a comfortable height with a kerf cut into it where the saw is inserted for filing. Post supports are driven into the ground at the ends of the saw.

SAW TEETH   Usually referred to the cutting teeth and not the raker or clearing teeth in the following relationships: (1) the tuttle tooth—two cutting, one clearing; (2) the lancet tooth—three cutting, one clearing; (3) the sterling tooth—four cutting, one clearing; (4) the great American or "M" tooth—three cutting, gap, three cutting. Until rakers were put in saws to pull out the sawdust, saws were not used for felling timber.

SAW TIMBER   Trees suitable for production of saw logs. Timber that would make merchantable lumber.

SAW VISE   Wooden or metal vise that held saws while being sharpened or set.

SAW WEDGE   Same as sawyer's wedge.

SAWYER   One who controlled the carriage and other machinery in sawing logs into lumber. The quantity and quality of the lumber depended on his judgement. Same as head sawyer.

SAWYERS   The men who fell trees and cut them into logs. In the central states and the western pine country.

—S—

SAWYER'S WEDGES   V-shaped, iron wedges inserted into the kerf when falling a tree to keep the tremendous weight of the tree from binding the saw.

SCALE   1. To ascertain the number of board feet in a log. 2. To keep a tally of total board feet of measure in logs, etc.

SCALE BOOK   A book especially designed for recording the contents of scaled logs.

SCALER or **LOG SCALER**   One who estimated the number of board feet in a log. His rule stick is said to be the cheat stick.

SCALE RULE   A long heavy hardwood ruler marked in inches and feet used in estimating the lumber in standing timber or in logs. It was usually four feet long but sometimes six. Same as cheat stick, log scale rule, money maker, robber's cane, thief stick, swindle stick.

SCALPER   1. One who sold lumber on commission. 2. Jacks who peeled hemlock bark.

SCANDIHOOVIAN DYNAMITE   Snoose, snuff.

SCANTLING   A piece of lumber of small size. Generally a two by four, or two by six, or smaller. Often called a two by scantling.

SCARF   Surface of the undercut in falling trees.

SCHNITZELBANK   A bench used in shaping shingles or shakes with a draw shave after blanks were cut from a block with a shingle froe. The wood pieces were held in a vicelike mechanism kept tight by the feet of the operator. A common machine in all logging camps.

SCHOOL MARM   A crotched log or tree with two main trunks. Since such logs did not turn easily in water, they were useful to raw river drivers in practicing the art of balancing on logs.

SCISSOR BILL   A worker indifferent to the interest of the laboring classes.

SCONE   Generally a rich baking powder biscuit containing currants or raisins.

SCOOP ROOF   Roof made of logs hollowed out and fitted to shed rain. Generally made of hollow basswood logs. Same as split roof.

SCOOT   1. A sled used to haul logs out of the woods by placing one end of the log on the sled. Same as crotch, drag sled, dray, bob, lizard, skidding sled, go-devil. 2. An inferior and practically worthless piece of hardwood lumber.

SCOOTER   Same as speeder.

SCORE   To hack timber with scoring axe before hewing to the line, as in hewing timbers.

SCORING AXE   A rather heavy (5–6 pounds) axe used to hack or score edges of logs before hewing to the line.

SCOW   A small river boat square-nosed at both ends.

SCRABBLE LOGS   Logs which came to the sorting jack without marks denoting ownership. Same as prize logs.

SCRAP   A popular (plug) chewing tobacco used in early logging camps. Same as scrap mixture.

SCRAPER   A dirt scraper. Same as slusher.

SCRAP MIXTURE   A cheap chewing tobacco that was sweet and juicy. Same as scrap.

SCRATCH BOY   A lad aout 12 years old who put scratch marks on each log as the scaler evaluated the logs at a landing.

SCRATCH STICK   A four foot long stick ending in a sharp, pointed iron to scratch log ends which had been evaluated by the scaler.

SCRATCHING AWL   A steel, sharp-pointed instrument used in marking wood.

SCREECH CAT   A high stump, usually from a windfall, which had splinters that whistled or sang in the wind. Sometimes wind blowing through the splinters made a hideous sound that scared lumberjacks.

SCREW JACK   Large jacks raised by a handle. Much similar to railroad jacks.

—S—

SCRIBNER   The Scribner rule. A log rule invented by J. M. Scribner in 1846. Today known as the Scribner decimal C.

SCRIP   Money issued by logging companies which could be used only at the company store.

SCROLL KNOB   Referred to one of the unique shapes of an axe handle. Specifically, the butt end of the axe handle.

SEAMS SQUIRRELS   Lice. Same as blue jackets, crumbs, gray backs.

SEASON or THE SEASON   1. The logging season, generally September 1 through April 1. 2. To age or dry lumber.

SEASON CHECK   A crack in timber caused by too rapid seasoning. Same as check.

SEASON'S CUT   The output of a sawmill for that portion of the year the mill was operated.

SECOND COOK   In a large camp, the second cook was the meat cook.

SECOND GROWTH   Young growth of timber coming up after a cutting operation or a forest fire.

SECOND LOADER   One who handled tongs that hooked into logs when cars were being loaded.

SECTION   A square mile (640 acres) of land. A survey unit into which the public lands of the United States are divided. There are thirty-six sections in a township.

SECTION CREW   Same as section gang.

SECTION GANG   A small group of workers who kept railroad lines in safe condition.

SECTION 37   1. Where all good loggers go when they "cash in their chips." No scalers are allowed. 2. There is no section 37; a township is divided into 36 sections.

SEE THE DENTIST   The excuse a lumberjack used to get away from camp and "live it up" in town for awhile.

SELECTIVE CUTTING   A method of cutting timber that took only selected trees from a stand. Usually applied to trees marked by a forester for removal under a forest management program. Not used in early logging, and later only by a few lumber companies. Probably the first selective logging was cutting the veneer logs. Actually, cutting the pine and leaving the hardwood was a form of selective logging. After the cutting of the big white pine, land so cut came to be the basis of the first great wave of tax delinquency in the 1870s and 1880s. The foundation of a new generation of timber barons was laid during this period when men saw a future in maple, yellow birch, basswood, oak, and hemlock and bought large blocks of tax certificates for very little money.

SELF-LOADING DAM   A dam in which long timbers were set on the upstream side at an angle to water surface. Same as rafter dam, slant dam.

SEND 'ER UP   Give her more steam at a steam jammer operation.

SENDER UPPER   Same as sender or send up man.

SENDER or SEND UP MAN   A member of the loading crew who attached the tongs to the logs and guided the logs up the skids. Same as bottom loader, ground loader, hooker, hooker on.

SENT HIM DOWN THE ROAD   To discharge a man.

SET   1. In saws, the degree to which the teeth were flared out so that the cutting edge would make a wider cut than the back of the saw, to prevent binding. 2. Two buckers or two fallers working together. 3. A device in which a saw was set to fix the correct set of the teeth or the correct flare to the teeth. 4. A logging camp. 5. A temporary site of a portable sawmill.

SET POLE   A long pole used in holding a bateau in place.

SETTER   The man who rode the carriage in a sawmill to adjust the saw to the proper cutting thickness as signaled by the head sawyer.

SETTING   The temporary station of a portable sawmill; a yarding engine. The ground within yarding distance of the spar tree.

—S—

SET UP   Seating arrangements at the mess tables in the cook shanty. Always arranged by the cook. Old-timers had first choice of seats; newcomers took any other places.

SETWORKS   Collective term for all the saw carriage parts.

SEVEN UP   A popular bunkhouse card game.

SHACKER   A man who lived, generally by himself, in a small building of logs or rough lumber.

SHACKING   Living in the woods, generally alone in a log building.

SHACKLE   A clevis.

SHACK UP   1. Getting out of the woods in the off season. 2. To take it easy, hole up waiting for better times.

SHAKE   1. A thick wood shingle made by splitting flat strips from a bolt. Generally made from cedar, sometimes pine, generally half an inch thick, six to twelve inches wide, and two to four feet long. 2. A crack in timber caused by frost or wind. Same as wind shake.

SHAKE BOLTS   Wood blocks from which shakes or shingles were made. Pieces removed first from blocks with a shingle froe and then shaped on a shingle bench.

SHAKE ROOF   A roof on camp buildings made of cedar or pine shingles or shakes which were split with a shingle froe.

SHAKE THE SIEVE   A term applied when men were to be laid off or the crew cut down in size.

SHAKES   Split strips of wood (usually cedar) laid down from eaves to ridgepole of early camp roofs used in place of shingles.

SHAKY LOG   The base of a tree which has been split by swaying in the wind.

SHAKY LUMBER   Lumber made from a shaky log.

SHANTY   1. Originally in the Canadian Gatineau, all the buildings of a camp complex. 2. Men's camp, men's sleeping quarters, bunkhouse. Sometimes, any small building. 3. A lumberjack. A term not used much in the lake states except in song and folklore.

SHANTY BOAT   The cook's raft which followed the log drive down the river. Same as wanigan.

SHANTY BOSS   A camp worker who got the wood and water and did chores. Same as chore boy.

SHANTY BOY   Original name for the lumberjack. Used mostly in song and verse. Original name for woods-workers—Used many years before the term "lumberjack."

SHANTY MAN   A lumberjack.

SHANTY NASTY   One who stayed on, living in one of the buildings of an abandoned logging camp.

SHARP-SHOD   Horses with shoes that have very sharp calks to enable the animals to walk on iced roads.

SHAVETAIL   A mule.

SHAVING CREW   The planers who smoothed the rough lumber.

SHAVING HORSE   Same as schnitzelbank.

SHAVINGS   The debris of a planing mill operation.

SHAVINGS VAULT   A large binlike building used to hold excess shavings from the planing mill.

SHAVING THE WHISKERS ON BIG DICK   Going from woods work to the wheat harvest in the Dakotas.

SHAW BABBITT METAL FORMULA   A babbitt metal formula used in bearings and journals consisting of six parts of black tin to one part of copper but no lead or antimony.

SHAY   A gear-driven locomotive on a logging railroad. Same as lima.

SHAY'S FOLLY   What jacks called the Shay geared locomotive.

SHEAVED WHEELS   Grooved metal wheels used on various types of trucks on a pole road.

—S—

SHEAVES   The grooved wheels in a block or pulley.

SHE-BOILERS   Women cooks.

SHEER BOOM   A heavy log boom placed diagonally across a river to divert logs into a sorting works or a holding boom to increase water flow. Same as fender boom, glancing boom.

SHEER SKID   A log placed on the lower side of a skidding trail on a slope to hold the log on the trail while being skidded. Same as fender skid, breastwork log, glancer.

SHEETING LUMBER   Fourth grade lumber, the lowest grade.

SHE PULLS   The downriver floating action of the logs after the dam was broken.

SHE'S A RAINBOW   What a day! *or*: A log with a severe kink in it.

SHE'S GOING TO HAUL   The log jam is about to break.

SHE'S NOTCHED   A tree into which the undercut is already chopped by axe prior to sawing.

SHIM   Blocking placed on crossties to level up railroad track; generally used on softwood ties, such as ceder. Also used to keep track from sinking into mud or to level logs or timbers in any type of construction.

SHIN CUTTER   An adze tool used in hewing logs to form the floors of stables and shanties. It had a handle long enough to enable a worker to use it while standing. The lumberjacks called it a shin cutter because if it slipped a man could cut his leg. Same as adze.

SHINGLE BENCH   A schnitzelbank or bench used for shaping shingles or shakes after blanks were cut from a block with a shingle froe.

SHINGLE BOLT   A short log, sometimes split, used as raw material to make shingles and shakes. Same as bolt, spoke bolt, stave bolt.

SHINGLE FROE   A tool for splitting shingles or shakes from a block of wood.

SHINGLE TOW   The log shaving produced while making shingles with a shingle mill.

SHINGLE WEAVER   A worker in a shingle mill.

SHINGLING LOGS   Sinking one end of a floating log by putting another log on top of it. A boom was put around a unit of these logs to transport them. They floated low in the water.

SHINNY   Intoxicating liquor.

SHIP AUGER   A long-shafted auger of various diameters used in dam and pier construction. Same as bridge auger.

SHIP KNEE   Large piece of wood taken from a tree at the point where the trunk merges into the roots, flattening out at this point. Piece almost forms a right angle and was used as support or brace in buildings and boats, or to make travois.

SHIP SAW   An early type of water sawmill.

SHIPWRIGHT AXE   A heavy, hewing axe used to shape ship's timbers.

SHOE CALKS   Sharp pointed spikes strapped to shoes for walking over ice.

SHOEING HAMMER or **FARRIER'S HAMMER**   Tool for driving the horseshoe nails into the horse's hoof.

SHOEING HARDY   A box mounted on four short feet and having small compartments in it to hold a blacksmith's horse shoe nails, hoof knife and the other tools used in shoeing a horse.

SHOEING SLING   A canvas or leather fastened to a horse or ox shoeing rack and used to lift the animal off balance in order to shoe it.

SHOEING RACK or **OX SHOEING RACK**   Stanchion-like racks in which oxen were lifted off balance in order to place two shoes on their divided hoofs.

SHOEING TREE or **SHOEING BUCK**   A rest for the camp blacksmith to use in shoeing horses. Made out of a natural growth of maple stump supported by roots. A horse rested his foot on the stump while being shod or having its hoof pared.

—S—

SHOE PACK PIE A pie made of vinegar, cornstarch, sugar, and water, sometimes flavored with lemon or vanilla.

SHOE PACKS Rubber bottoms with leather tops worn in cold weather. Worn with several pairs of socks. Same as packs.

SHOES The tires on sled runners. Usually made of cast iron on tote sleds, and thinner steel on logging sleighs.

SHOO FLY REAR Taking only the logs that were easy to get in sacking the rear.

SHOO FLY SACKING Not sacking thoroughly. Not getting all of the logs thrown up on the river bank back into the river during the river drive or following a log jam.

SHOOKS Barrel staves, hoops and heads. Were shipped "knocked down" to save space.

SHOOT A JAM To loosen a log jam with dynamite.

SHOOT THE BEANS Pass the beans at the dinner table.

SHORE HOLD The attachment of the hawser of the raft of logs to an object on the shore.

SHORELINE An imaginary tool used to fool greenhorns.

SHORE TIMBER Timber that could be skidded directly to the river bank or that required only a very short sleigh haul.

SHORE UP Simply to reinforce anything—trestles, piers, bridges, excavations.

SHORT HAUL Referred to the overall distance a load of logs had to be moved to the landing.

SHORT ROAD A road upon which unloaded logging sleighs can return for reloading without meeting a loaded sleigh. Same as go back road.

SHORT STAKER A very migratory worker. One who stayed only a short time at any one camp.

SHORT STUFF   All bolt length (100 inches) timber including shingle bolts, tie cuts and pulpwood sticks.

SHORT SWEET'NIN'   A lump of sugar to sweeten coffee.

SHORT TONGUE   Used between bobs in place of cross chains, to hold the rear bob at its distance.

SHOT HIS WAD   Spend all his money on a "fling" in town.

SHOT GUN BUNKS   Old-fashioned bunks into which the lumberjack crawled from the foot of the bed. Same as muzzle loaders.

SHOTGUN FEED   The method of running a sawmill carriage by steam utilizing a long piston within a long cylinder resembling a shotgun.

SHOULDER ROADS   Narrow, iced sleigh roads (54 inches between runners) with no ruts but with iced shoulders for the outside edges of the runners. Same as troughed roads.

SHOVE HIS NOSE DOWN   To thrash or whip a person.

SHOW   A logging operation or a logging chance. Spoken of as a good show or a poor show.

SHOW HIM THE TOTE ROAD   Fire him!

SHOWING THE WHITE FEATHER   Said of a river pig or other woods worker who stopped work because of bad weather.

SHRINKAGE   Generally referred to log loss during a river drive; lumber shrinks in drying.

SHROUD   A blanket.

SHUFFLE THE BROGUE or **HUNT THE SLIPPER**   An old bunkhouse game.

SHUT THE HOLE   A river driver going out of sight after falling into the river.

SIDE   One complete yarding and loading crew, with spar tree and donkey engines. A two-side camp has two spars and crews.

—S—

SIDE BOOM   Boom stretched along the shore to prevent logs from becoming stranded.

SIDE CUTS   Meanderings of a stream parallel to one another.

SIDE DELIVERY BUNK   Bunks, built parallel to the deacon seat, into which the lumberjacks crawled from the side, in contrast to muzzle-loaders which he entered from the end.

SIDE DOOR PULLMANS   Boxcars on the railroad.

SIDING MILL   A re-saw mill at the sawmill which cuts boards or planks into two or more boards of desired thickness.

SIDE JAM   A jam which had formed on one side of a stream, usually where the logs were forced to the shore at a bend by the current, or where the water was shallow or rocks were partially submerged. Same as jam, log jam, plug.

SIDE JAMMER   Same as swing jammer.

SIDE MARK   A log ownership mark placed within a blazed area on the side with a hammer or scribe or other cutting blade. These side marks were often easier to read in the water at sorting booms than were end marks. Same as log brand, log mark.

SIDE PUSH   A boss in a logging camp. Same as head push, straw boss, working foreman.

SIDEWHEELERS   Lice, crumbs, livestock.

SIDEWINDER   1. A big spree. 2. A falling tree, which, hitting another, rolls on its axis. 3. A tree knocked down by the fall of another tree. 4. Logging locomotive with four or five cylinders on a side with the rods going vertically instead of horizontally.

SIGHTS   A timber cruiser's compass bearing.

SILKALEEN TENTS   Double tents with an air space between used by timber cruisers and land lookers in winter.

SILVER CUP   A popular (plug) chewing tobacco used in early logging camps.

SIMMONS CEDAR KING   A popular cedar saw having no raker teeth.

SINGLE-BITTED AXE   A poll axe with one cutting edge.

SINGLE COUPLER   Similar to a slidding grab but with a very short chain.

SINGLE-CUTTING BAND SAW   A band saw with teeth only on one side.

SINGLE LINE   A long log chain used to send up logs when loading a sleigh without a top loader.

SINGLETREE   Bar to which the traces were attached to one horse in skidding logs.

SINKAGE   Loss of logs being driven downstream to mills causing shrinkage.

SINKER BOAT   A raft carrying a windlass and grapple, used to recover sunken logs. Same as catamaran, monitor, pontoon.

SINKERS   1. Doughnuts, biscuits. Same as washer, stovelids, string of flats. 2. Water-soaked logs lying on the stream or lake bottom. Same as dead head or bobber.

SINK HOLES   Railroad tracks laid over bog or swamp areas would develop spots where tracks would sink out of sight.

SIOUX PANTS   Trousers of heavy wool having a very long nap.

SIWASH   To drag logs up to the main hauling cable.

SIWASHING   Tangling of a line on a stump or other obstruction.

SIZZLER   A cook who generally fried food. See boiler.

SKELETON CAR   The Russell railroad car was a skeleton car since it had no deck, only bunks.

SKELETON TRACKS   Logging spur lines used only in winter with ties and rails laid on frozen ground. No ballast used.

SKID   1. To drag logs from the place they are cut to the skidway, landing or mill. Same as drag in, dray in, snake, twitch. 2. A log or pole, commonly used in pairs, upon which logs were handled or piled. 3. The log or pole laid in a skid road to reinforce it. 4. A piece of hardwood about six feet long with studs on one side and two hooks on one end. It was placed on edge of dray to roll logs onto dray with a cant hook. The studs kept the log from slipping back.

SKIDDED   Drunk and unconscious because too much whiskey.

SKIDDER   1. A teamster skidding logs out of the woods. 2. A yarding and loading engine which has a steel tower in place of a spar tree.

SKIDDING CHAIN   Chain usually sixteen feet long with round hook on each end to attach to log for skidding or dragging. Same as bunching chain.

SKIDDING DRAY   A one-bunk dray used only for skidding.

SKIDDING GRAB   A short chain with two chisel hooks at each end and a bitch link in the center. Used in horse skidding.

SKIDDING HOOKS   A large pair of hooks attached by links to a ring used for grappling logs in skidding or loading. Same as skidding tongs, chain grapples.

SKIDDING SKINNER   Skidding teamster.

SKIDDING SLED   Two runners with bunk in center to haul logs out of the woods. Same as dray.

SKIDDING TEAM   The team used in dragging logs out of the woods.

SKIDDING TONGS   A pair of hooks attached by links to a ring used for grabbing and holding logs in skidding or loading. Same as chain grapples, coupling grab, dogs, grapples, grabs, grips, skidding hooks.

SKIDDING TRAIL   A road over which oxen or horses pulled logs. Same as skid road, gutter road, snake trail.

SKID GREASE   1. Butter. 2. A heavy oil applied to skids to lessen the friction of logs dragged over them.

SKID POLES   Four to six inch thick poles set from the ground to a sleigh or railroad car to form a support up which logs were decked in a X-haul.

SKID ROAD   1. A road over which oxen, horses, or tractors pulled logs. Generally a short wide road rather than a main road. Same as gutter road, skidding trail, snake trail. 2. That part of a city where loggers congregated when in town. (Generally known as skid row.)

SKIDS   Same as skid timbers.

SKID TIMBERS   Large timbers used at the landing to hold and support logs before loading.

SKID UP   1. To level or reinforce a logging road by the use of skids. 2. To collect logs and pile them on a skidway.

SKIDWAY   1. A road along which logs were piled. 2. Sometimes used for a pile of logs waiting to be loaded. 3. Location where the logs were hauled across skids and rolled forward to load the logging sleigh.

SKIMMING OFF THE CREAM   Originally, cutting only the choicest white pine near the drive river banks with logs at least eighteen inches in diameter.

SKIN LOGS   De-bark logs or peel logs.

SKINNER   A teamster of oxen or horses. A bull skinner when applied to oxen.

SKIP GRAB   An automatic locking and unlocking device used on downhill skidways. Logs out of control would skid down the hill when the skip grab snapped loose and the horses had stepped aside.

SKIP or SKIPPER   1. A hammer used to drive a dog into a log. One end was a sledge, and the other could be used to drive a hook out of a log again. 2. A foreman; a boss.

SKIRT   A woman.

SKY BIRD   A top loader. Same as sky hooker.

SKYBOUND   A tree that refuses to fall even when wedged.

SKY HOOK   Mythical, all-powerful hook. One end of a line fastened to the sky, which hook tenders cried for when they had to fight against bad hang ups.

—S—

SKY HOOKER   A member of the loading crew who stood on the logs as they were sent up. Same as sky bird, top loader.

SKY HOOKING   The work of arranging the logs on top of the car or sleigh.

SKY LINE   A logging method where a taut line reaches between two spar trees.

SKY LOADER   Same as sky hooker.

SKY MAN   Same as top loader.

SKY PIECE   A hat or cap. A top loader.

SKY PILOT   A priest or minister who went from camp to camp conducting the services and often remained after supper to lead hymn singing in the bunkhouse. Also called in when lumberjack was about to die.

SKY PILOT'S ORGAN   A small, portable organ carried with sky pilots as they made their rounds of the logging camps.

SKY ROUTE   The way to Heaven.

SLABBED   A damaged log; piece or chunk broken off in skidding, loading, felling, or bucking. Careless work.

SLABBER GATE   A gate saw used for sawing slabs.

SLAB MAN   Same as slab picker.

SLAB PICKERS   Sawmill workers who removed slabs in the grading process.

SLAB SAW   Saw that removes the outer slabs or bark edges from saw logs, exposing the body wood. Same as gate saw.

SLAB WOOD   Wood from edgings and slabs from the sawmill as distinguished from body wood.

SLACK WATER   The temporary slackening of the current caused by the formation of a jam during a river drive.

SLANT DAM   A dam in which long timbers were set on the upstream side at an angle of twenty to forty degrees to the water's surface. Same as rafter dam, self-loading dam.

SLASH   The limbs, tops, and unused logs left after a tract of land had been logged. It was like tinder when forest fires started. Same as slashing.

SLASHBOARD   A dropleaf in a dam.

SLASHER   1. Man who cleaned up the slash after a logging operation. 2. A machine fitted with one or more coarse circular saws for sawing lumber out of slabs.

SLASHINGS   The area of recently cut timber. A land area full of slash.

SLASH SAW   An "up and down" saw in the early water sawmills.

SLATS   A man's ribs.

SLAVE   Any wage earner.

SLAVE DRIVER   A foreman, especially a strict one.

SLAVE MARKET   An employment office.

SLED   A logging sled used to haul logs on snow or ice. Called a sleigh in the lake states. Bobs, skidding sleds, drays, log boats, lizards, joe-hog, go-devil, travois, mud boat, scoot.

SLED KNEE   A wrought-iron casting which fastens the sled runner to the bed of a logging sleigh.

SLED ROAD   Same as sleigh road.

SLED TENDER   1. One who assisted in loading and unloading logs or skidding with a dray. Same as chain tender. 2. A member of the hauling crew who accompanied the logs to the landing, unhooked the grabs, and returned to the woods.

SLEDGES   Same as sleighs.

SLEEPING SHANTY   Sleeping quarters of the lumberjacks. Same as bunkhouse.

━S━

SLEEPING TENT   A tent set up on a river bank for the river drivers during a log drive. It held from eight to twenty men, sleeping on balsam boughs laid on the ground.

SLEEPING WANIGAN   A houseboat used solely for sleeping during the river drive.

SLEIGH   A logging sled for hauling logs on snow and ice. Called a sleigh west of Michigan. While sleigh and sled were used interchangeably, the lumberjack more often referred to it as a sleigh. Same as logging sled, sled, twin sleds, two sleds, wagon sleds.

SLEIGH DOGS   Snubbers or brakes used to slow down loaded logging sleighs on downhill runs. See snubber, downhill clevis and uphill clevis.

SLEIGH-HAUL CAMPS   Camps where sleighs were used to haul logs several miles to landings or banking grounds.

SLEIGH IRON   All the various pins, bolts, rods, plates, washers, and other iron parts used in construction of logging sleighs.

SLEIGH POLE   Same as sleigh tongue.

SLEIGH ROAD   A snow plow-like divice pulled by horses to clear skidways and tote roads of excessive snow.

SLEIGH ROLL BAR   A round, wood bar attached between the lead ends of logging sleigh runners and permanently connected to the tongue of the sleigh.

SLEIGH SHOES   Heavy, iron, curved facings on the bottom of sleigh runners.

SLEIGH TONGUE   A heavy pole attached to the roll bar orf a logging sleigh to which the horses' harnesses were attached.

SLEPT IN   Sundays were days of rest in camps. Jacks rose about the same time but did not go into the woods to work.

SLICK   A razor-sharp chisel, two to four inches wide. Blade had a handle twenty-five inches to thirty-five inches long. Also called a paring chisel.

SLICK AS A CHICKEN'S TEAT   An expression to denote something that was very slick.

SLIDE   1. A trough, usually made of logs, for moving logs down a slope by gravity. Same as dry slide. 2. A derrick for loading logs onto cars.

SLIDE ASS JAMMER   A jammer that pulled itself along.

SLIDE JAMMER   Same as slide ass jammer.

SLIDE TENDER   One who keeps a slide in repair.

SLIME   Sap of the hemlock tree which would adhere to the mitten and clothes of the barkers. It was a sticky nuisance hence the term slime.

SLING   A device enabling stream log tugs to be lifted out of the water for inspection and repairs.

SLIP   1. The incline plane that moved logs into sawmill. Same as gangway, jack ladder, log jack, log way. 2. A slusher.

SLIP GRAB   A pear-shaped link attached by a swivel to a skidding evener or whiffletree, through which the skidding chain was passed.

SLIP HOOK   A hook attached to a chain which was wrapped around a log being loaded by a jammer. Same as trip hook.

SLIPPER   1. A bar used to couple two logging sleighs. Same as goose neck, dray hook, bumper pole, rooster. 2. Jacks who peeled hemlock bark.

SLIP SKIDS   Freshly peeled skids on which logs slide instead of roll when loaded. Same as glisse skids.

SLIVER CAT   1. Splinter on stumps standing up two to four feet which whistle in the winter wind. 2. A splinter, usually about a yard long, attached to a pole by a piece of leather. The lumberjacks twirled this splinter by swinging the pole rapidly to make a howling, screaming sound that frightened the greenhorns. Often it was used while icing roads at night.

—S—

SLOAN'S LINIMENT   A favorite liniment in camps. Sometimes referred to bad whiskey.

SLOOP   Two runners with a bunk to haul logs out of the woods. Same as dray.

SLOOP LOGS   To haul logs down steep slopes on a dray or sloop equipped with a tongue.

SLOUGH PIG   Usually a second-rate river driver assigned to picking logs out of sloughs in advance of the rear.

SLUICING   Guiding logs through the gate of the dam.

SLUICE CREW   Crew that steered logs into the various pockets in a sorting boom.

SLUICED   1. Driving logs over a splash dam. 2. Upset, said of a man who has accidentally gone off the road and piled up.

SLUICED HORSES   Horses injured by loaded sleigh piling up on them on a steep hill because of improper braking of sleigh.

SLUICE DAM   A splash dam. A dam built to store a head of water for driving logs.

SLUICE GATE   The gate closing a sluiceway in a splash dam.

SLUICEWAY   Gate in a dam through which logs pass when they are floating downstream.

SLUM GULLION   A stew made of meat and vegetables.

SLUSH   Coffee, also know as mud.

SLUSH COOK   An assistant cook. Same as cookee, taffle.

SLUSHER   Horse-pulled earth scoop with twin handles manipulated by the teamster on road and railroad right-of-way construction work.

SLUSH HANDLER   A cookee. Same as slush cook.

SMALL GAME IN BUNKS   Bedbugs.

SMALL ROLL   One of the several small gates in a logging dam.

SMALL TIMBER   Trees left to cut after the main cut was made which included trees for ties, poles, posts, etc.

SMEAR   A bunkhouse card game.

SMILO   Hard liquor

SMOKE EATER   A forest fire fighter.

SMOKE HOLE   A large, square hole in the roofs of the early camp cambooses, directly over the fireplace below, that permitted the smoke of the open fire to escape.

SMOOTH AS A SCHOOL MARM'S THIGH   A phrase lumberjacks used for an exceptionally smooth board or other smooth object.

SNAG   1. A standing dead tree. Same as ram pike. 2. A sunken log or a submerged stump.

SNAKE or SNAKING   To horse drag a log to its destination without loading on a sleigh. Twitch skidding by chain only.

SNAKE HUNTER   A person who cam to a lumber camp wearing city clothes and pointed-toes shoes.

SNAKE ROOM   A room off a saloon, usually two or three steps down, into which a barkeeper or bouncer could slide drunk lumberjacks headfirst through swinging doors from the barroom.

SNAKE TRAIL   A skid road.

SNAP JACK   A river pig.

SNAP LINE   1. A rope or cable brake used to ease double headers downhill. 2. A line to keep machinery erect. Same as snub.

SNATCH BLOCK   A metal pulley with a side release plate for the cable or chain.

SNATCHER   A man on a lumber raft who manipulated the snatch pole.

SNATCH POLE   A long pole on lumber raft to lift the front raft out of water in going over falls. Also used to hold onto in rough water. Also used to snub a raft on sharp bends in river.

— S —

SNATCH TEAM An extra team stationed at an incline in a logging road to assist in pulling a loaded sleigh up a hill. Same as tow team.

SNIB In river driving, to be carried away purposely, but ostensibly by accident, on the first portion of a jam that moves; to ride away from work under the guise of being accidentally carried off.

SNIPE To round off the end of a log in order to make it drag or slip more easily. Same as nose.

SNIPED Same as nosed.

SNIPER One who rounds off the end of a log before it is skidded.

SNOOSE or SNUSS Damp snuff for chewing, placed under upper or lower lip. It is said that no one knows what was in it, but all agree it was the most powerful chewing tobacco. Same as Copenhagen, Scandihoovian dynamite, Swedish brain food, Swedish conditioner powder, galloping dust.

SNORT or SNORTER A drink of whiskey.

SNORTING POLE A sapling or a cant hook handle which was laid under the hay in a muzzle loader to separate the sleepers.

SNOW A ROAD To cover bare spots in a logging road with snow to facilitate the passage of sleighs.

SNOWBALL HAMMER Same as snow knocker.

SNOW-ICED ROADS Entire road iced with no iced ruts.

SNOW KNOCKER Small hammer with one end of the head a pick. Clipped onto the harness or sleigh, it was used to knock or chip ice or packed snow from the hoofs of horses. Same as ball hammer or peen hammer.

SNOW ROAD DRAG A snowplow-like device pulled by horses to clear skidways and tote roads of excessive snow.

SNOW ROADS Roads made of packed snow. Often rolled with a snow roller to firm up and harden the snow.

SNOW ROLLER   Huge, heavy roller used to pack down snow on roads. Often the snow was packed by the feet of the lumberjacks.

SNOW SLIDE   A temporary slide on a steep slope, made by dragging a large log through deep snow when it was soft or thawing; when frozen solid it could be used to slide logs to a point where they could be reached by sleighs.

SNOW SNAKE   A travois. Sometimes used in reference to steam hauler.

SNOW TUNNELS   During winters of heavy snowfall, deep pathways called "snow tunnels" were dug out between camp buildings and the railroad logging spurs running into the camp complex.

SNUB   1. A line to ease sleighs down a hill. Same as snap line. 2. To check, usually by means of line around a tree or stump, the speed of logging sleds or logs on steep slopes, or of a log raft.

SNUBBER   1. A device for checking the speed of a sleigh down a steep hill by winding a rope around a stump and letting it out slowly. 2. A device with a drum and brake, used with a cable for lowering logs or cars down a steep grade.

SNUB LINE   Same as snubber.

S.O.B.   The situation in which a log or logs are so placed that they squeeze out those below them and tumble the pile. Also used in many, many other situations.

SO BIG IT TAKES THREE DAYS TO WALK AROUND IT   An expression describing anything big.

SOCKET PEAVEY   A peavey with a handle fitted into the socket which is armed at the lower end with a pike, and on the upper end of the socket is a clasp to which the bolt is bolted.

SOFT LADIES   Whores.

SOFTWOOD   A needle-leafed, or coniferous tree, as opposed to the broadleaf or hardwood tree. The pines are softwood.

SOLDIER   To evade work; to loaf.

206

—S—

SOLID JAM   1. In river driving, a jam which formed solidly and extended from bank to bank of a stream. 2. A drive was said to be in a solid jam when the stream was full of logs from the rear to the mill, sorting jack, or storage boom.

SORTER   Man who worked at a booming grounds. Same as culler.

SORTING BAY   Boomed off area in lake or stream where logs were sorted for size, species, or grade.

SORTING BOOM   A strong boom used to guide logs into the sorting jack.

SORTING CAMP   A rivermen's camp located near the sorting works.

SORTING GROUNDS   Booming area where logs were sorted for any reason.

SORTING JACK or **SORTING GAP**   A raft, secured in a stream, with an opening through which logs passed to be sorted by their marks and diverted into pocket booms of the downstream channel.

SORTING PEN   Same as sorting pocket.

SORTING POCKET   A division or section of a sorting grounds where logs were held after having been taken out of a drive or boom.

SORTING SHED   A building at the sawmill where sawed lumber was graded as to species, quality, and dimensions.

SORTING TABLES   Wide, flat benches where lumber is sorted and graded in a sawmill.

SORTING WORKS   An enclosure on a river where logs were sorted as to ownership. Same as boom works.

SOUGAN or **SOOGAN**   A heavy blanket; name applied generally in the early days when the lumberjacks carried their own bedrolls.

SOUND   To pound a standing tree with the flat side of an axe to see if it was sound, that is, not rotted in the core.

SOUND KNOT   A knot which is solid across its face, as hard as the surrounding wood, and so fixed that it will retain its place in the piece.

SOUP SHANTIES   1. Buildings where employees of boom companies were boarded near the booming grounds. 2. Sometimes, the cook shack or shanty.

SOUR DOUGH COOK   A poor bread maker.

SOW BELLY   Salt pork. Much of the pork was salted to preserve it. Generally referred to the pork side or bacon portion.

SOW BOSOM   Bacon.

SPACING BAR   Bar used for spacing logs when starting a turn of logs.

SPANISH WINDLASS   A device for moving heavy objects in logging. It consisted of a rope or chain, within a turn of which a lever was inserted and power was gained by twisting. Same as twister.

SPAR TIMBER   The large, round timber used for the masts of ships.

SPAR TREE   Same as high-lead tree.

SPARK CAP or **SPARK CATCHER**   A spark arrester on a steam engine or locomotive.

SPEARHEAD   A popular smoking or chewing tobacco.

SPEEDER   A gasoline-driven railroad hand car.

SPIDER   Heavy cast-iron frying pan.

SPIDER WEB   Log boom strung across the mouth of feeder stream to catch escaped logs.

SPIKE   To add snoose to chewing tobacco.

SPIKE KNOT   A knot sawed in a lengthwise direction.

SPIKED SKID   A skid in which spikes were inserted in order to keep logs from sliding back when being loaded or piled. Generally called skid.

—S—

SPIKER A lumberjack who has quit his job. Same as hiker.

SPIKE MAUL A heavy sledge-like hammer used to drive railroad spikes.

SPILE A small log used as a pile in dam construction.

SPILING A series of spiles driven in front of a dam.

SPILL 1. A log dump. An overturned load of logs was also called a spill. 2. To let the water through a dam.

SPILLWAY The overflow of water over dam in stream or lake.

SPIRAL AUGER An auger with a down-cutting lip and an open spiral shank.

SPLASH To drive logs by releasing a head of water confined by a splash dam. Same as flood.

SPLASH BOARDS Boards placed temporarily on top of a dam to heighten the dam and thus increase the head of water available for river driving.

SPLASH DAM A dam built to store a head of water for driving logs. Same as flood dam, squirt dam, flash dam.

SPLASHES Water released from flooding dams at intervals to assist in driving logs downstream.

SPLINTER BELLY A migratory carpenter.

SPLIT-LOG FLOOR Same as puncheon floor.

SPLIT ROOF or **SPLIT POLE ROOF** A roof of a logging camp or barn made by laying strips from straight-grained timber. The strips run from the ridge pole to the eaves; joints are broken with other strips, as in a shingle roof. Same as scoop roof.

SPLIT SHINGLES Shingles cut by hand with a shingle froe.

SPLITTER Man who made the top cut in a hemlock barking crew.

SPLITTING SAW A saw used to cut shingle blocks into quarters which were then fed into the shingle mill.

SPLITTING WEDGE   A heavy, wide V-shaped, metal wedge used in splitting rails or making firewood.

SPOKE BOLT   A short log sometimes split. Used as primary raw material for making spokes for wheels. Same as bolt.

SPOOL   Same as capstan or a drum on a steam donkey engine.

SPOOL BARS   Horizontal levers attached to a capstan.

SPOON or FLOP   Let's turn over together. A term used in the earliest and crudest camps when men were crowded together to keep warm while sleeping.

SPORTING HOUSE   A dive, house of ill fame.

SPOT   1. To place logging cars in position for loading. 2. To put a blaze on a tree.

SPOTTING OUT AREA   An area where loaded sleighs were left at the end of a work day to be ready for continued hauling in early morning of the next day.

SPREAD   A very long blanket used in the early camps to cover several men.

SPREADER   A wooden hitching device that held apart a crotched chain used in skidding in place of doubletrees.

SPREADER CHAIN   Chains attached to the spreader in a skidding harness.

SPREADER RINGS   Brightly-colored, celluloid rings as decorative parts of a horse harness.

SPRINGBOARD   A narrow platform on which fallers stand while falling a tree.

SPRING BREAK UP   The time when logging camps closed for the season because transportation became too difficult due to thawing of snow and ice in the woods. Also, the time when the ice in the river broke up and floated downstream so that the drive could start.

—S—

SPRING CUT   Should a camp fail to achieve its desired objective of logs in fall and winter, a spring cut was made and the logs skidded directly to the river.

SPRING POLE   A springy pole attached to the tongue of a logging sled for holding the weight of the tongue off the horses' necks.

SPRING POLES   A sleigh pole was too heavy to be supported by the horses' necks. To support the sleigh pole, a metal rod ran along the top of the pole and was attached to a green ash pole known as the spring pole which fitted into the two rings attached to the front part of the sleigh beam. The spring poles were replaced each year to assure "spring" in the poles.

SPRING RISE   High water on a dammed stream.

SPRINKLER   A large watertight box built on a sleigh. It sprinkled the logging road with water to make an ice road for sleigh haul. It held about 500 gallons of water and was used at night when water froze better, or on Sunday. Same as ice box, icer, ice tank, tank, water box, water tank, water wagon, water cart. Lumberjacks used to say that a sprinkler was like a drink of whiskey—when you needed it, you needed it awful bad.

SPRINKLER SLEDS   The sleds upon which the sprinkler was mounted. Two sleds, whose runners turned up at both ends, were fastened together by cross chains with a pole at each end, so that the sprinkler, or water tank, could be hauled in either direction without turning around.

SPRINKLING   Same as tanking.

SPRUCE GUM   Sticky, sap balls exuded by the spruce tree collecting in more or less ball-shaped masses on the outer bark of the tree. Often chewed by the early loggers in place of tobacco.

SPRUCE GUM BOXES   A small, wooden box with sliding cover whittled by jacks in the bunkhouse. Made to hold spruce gum balls and given to sweethearts and wives "back home."

SPUD   1. A tool for peeling the bark off logs, generally hemlock bark. Same as bark spud, barking iron. 2. A potato.

SPUDDER A man who removed the bark in a hemlock barking crew. Same as barker, peeler.

SPUDDING Removing the bark of logs with barking spuds.

SPUDDING IRONS Same as barking irons or barking spuds.

SPUR or SPUR LINE A branch logging road, either a sleigh road or a railroad.

SQUARE AWAY 1. To get tools and gear all set to go to work. 2. Setting logs parallel to a haul road or tracks preparatory to loading.

SQUARE LOAD Logs piled in a square-shapped load on a logging sleigh.

SQUARED OFF Same as sniped.

SQUARED TIMBER Cants or hewn timbers.

SQUARING CORD Chalk line used in hewing logs. A marking cord.

SQUARING THE BUTT Cutting off the ragged ends on a butt log.

SQUAW CANT The clip on the cant hook.

SQUAW DANCE Same as stag dance.

SQUAW LINK A double-ended grab link used to connect the loose end of a broken chain as temporary or emergency repair.

SQUAW WOOD The butt wood with much resin in it was traded by jacks at local whore houses for "services." This wood was then used by the "Madame" to start the stove fires in the whore house. Same as "trading stock."

SQUEAK HEEL Chilblains, sore heel from wet and chilled feet.

SQUEEGEE A plunger, made from a large tin can and a stick, for washing clothes in a tub or a large can of water.

SQUEEZE BOX A concertina or accordian. One of the musical instruments jacks played for bunkhouse entertainment after supper.

SQUINT EYES The camp saw filer.

—S—

SQUIRREL   Roughhouse game in the bunkhouse.

SQUIRREL WHISKEY   A cheap brand sold by early saloon keepers. When drunk as an eye-opener before breakfast, it was supposed to make a man climb a tree. Mule.

SQUIRT DAM   Dam usually built on a small tributary to the driving stream which, when opened, gave extra water. Same as splash dam, flash dam, flood dam.

ST. CROIX SCALE   A method of measuring board feet of lumber.

STACK   1. A pile of pancakes. 2. A pile of lumber or logs. 3. Smokestack on a donkey engine or woods locomotive.

STACKING YARD   Area at a sawmill where green freshly cut lumber was stacked.

STAG DANCE   Jacks danced among themselves to the tunes of the camp fiddler. The "ladies" wore a red hankie around one arm or tied a sack around themselves.

STAGE   Referred to depth of water and its relationship to a log drive.

STAGE OF WATER   The amount of water in the river for driving logs. In a good stage, the banks were full of water. If too high, the logs would catch on obstructions; if too low, the logs would not float.

STAGGING TROUSERS   To cut trousers off at the bottom for easier working in snow and on river drives.

STAGS   Heavy pants cut off at the bottom of the legs about at boot tops.

STAG SHOES   Shoes with tops cut off to form slippers.

STAG TREE   Tree whose top has been broken off.

STAKE   1. Wages earned on a job. 2. In a timber survey, a corner post.

STAKE BOUND   Describing one who had wages enough due him to make him restless to get to town. Same as stakey logger.

STAKE POCKET   A three-sided, iron pocket affixed to logging sleighs and railroad flat cars into which hardwood stakes were inserted to hold logs in place during transit.

STAKEY   Jack with lots of money.

STAKEY LOGGER   Describing one who had wages enough due him to make him restless to get to town. Same as stake bound.

STAMP   Mark on log showing ownership.

STAMP AXE   A hammer with raised letters used to stamp ownership identification on logs. It was made from a poll axe with the letters on the pole side. Same as stamping hammer.

STAMPER   A log marker or stamping hammer.

STAMPING HAMMER   A hammer with raised letters or numbers which were stamped on logs to indicate ownership. Used when logs belonging to more than one firm floated down the river in the same log drive. Same as branding axe, log stamper, marking hammer, marking iron, stamping axe, or branding hammer.

STAMPING IRON   A bar with raised letters on end for marking logs. Same as marking iron.

STAMP MARKER   Man who used the log-marking hammer to denote log ownership.

STAMP MARKS   Marks left in log ends by the use of a stamp hammer (Log-marking hammer).

STAMPS   Short for stamp hammer.

STAND   1. Standing timber. 2. Any piece of timberland.

STANDARD   A very "strong" popular smoking and chewing tobacco in camps used by jacks. Also Peerless, Climax, and Spearhead.

—S—

STANDARD GAUGE   A railroad track four feet eight and a half inches wide; compare narrow gauge track.

STAND CLEAR   Woods cry to watch out for the tree being felled. Same as cry "t-i-m-b-e-r."

STANDING BOOM   A permanent set of boom sticks framing rectangular log raft assembling grounds. Pilings hold these heavy boom sticks in place.

STAR CHIEF   A camp cook.

STAR LIGHT   The skylight in the roof of the sleeping shanty.

STAR LOAD   Same as prize load.

STARTING BAR   1. A heavy bar used to jar loose a sleigh which had frozen to the icy road. It was usually used as a lever to raise the runner. 2. Used to move railroad cars short distances.

STARTS   The heavy rod driven into the sleigh runner, through the rave, to hold the beam in place.

STATE OF MAINE SHANTY   Same as camboose.

STATIONARY ENGINE   Used for skidding or uphill hauls. Same as donkey engine.

STATION MEN   Men who built logging roads or railroads for so much a lineal unit.

STATIONS   Segments of 100 feet of railroad track. Men who laid this track were paid on the basis of these segments.

STAVE BARROW   A wheelbarrow box made of barrel staves.

STAVE BOLT   A short log sometimes split. Used as primary raw material for making spokes for wheels. Same as bolt.

STAVE BOLTS   Wood to be sawed into barrel staves.

STAVE MILL   Mill in which barrel staves were cut and shaped.

STAVES   Narrow hardwood strips forming the sides of barrels.

STAY BOOM   A boom fastened to a main boom and attached upstream to the shore to give added strength to the main boom.

STEAM CAT   Steam hauler.

STEAM DAGOS   Drag saws driven by compressed air.

STEAM FEED   Refers to the power used to move logs back and forth against the saws in a sawmill.

STEAM HAULER   A wood-burning steam engine, the forerunner of the modern tractor; used to haul trains or sleighs of loaded logs over iced roads. It had caterpillar rear wheels with sleigh runners in front. The steam hauler would pull ten or more sleighs grouped into trains with each sleigh loaded with 10,000 to 15,000 board feet of logs. This meant that from 150,000 to 100,000 board feet of logs were hauled each trip.

STEAM JAMMER   A machine operated by steam engine and used for loading logs upon cars. Same as loader, steam loader.

STEAM KICKERS   Steam-driven, stout, steel pins used to spin logs over in a sawmill.

STEAM LOADER   Large, steam loading machines usually mounted on a railroad car. A steam jammer.

STEAM POWER SAWMILL   Sawmill wherein saws were driven by steam power.

STEAM SAW   A steam-driven, portable saw used in later-day logging operations in the lake states.

STEAM SKIDDER   In later logging years, any steam-driven vehicle used to skid logs to hauling roads. A donkey engine.

STEAM SPRINKLER   A heavy sleigh with a large steam boiler mounted on it. Steam was generated and piped into the sleigh runners and sprayed into the ice ruts to re-form ruts through which loaded logging sleighs were pulled.

STEAM UP   To fire up a steam locomotive or a donkey engine preparatory to getting started at the day's work.

—S—

STEEL GANG   Railroad track-laying crew.

STEERING OAR   A very long heavy oar used to steer a bateau or lumber and log rafts.

STEERING SWEEP   Same as steering oar.

STEERSMAN   Men who steered lumber and log rafts downriver.

STEM   1. A lumberjack's leg. 2. The main street in town.

STEMWINDER   A geared locomotive such as the shay, climax, etc.

STERNO STIFF   Drinker of denatured alcohol.

STEW   A toot; to get drunk.

STEW BUILDER   A cook in a logging camp. A gut robber.

STEW BUM   A camp cook.

STICKER   A hung up lumber or log raft.

STICKS   100 inch pulpwood logs or squared timber.

STIFF   Any working man not a white-collar or office worker.

STIFF TUG ROAD   An iced road not level but just slightly uphill. So called because there was never any slack in the harness tugs.

STILL WATER   That part of a stream having such slight fall that no current is apparent. This caused problems in log drives. Same as dead water.

STITCHING HORSE   Same as harness vise.

STOMACH ROBBER   The cook in a logging camp.

STONEBOATS   Heavy, wooden sled-like devices used in hauling stones from fields or roads.

STORAGE BOOM   A strong boom used to hold floating logs in storage at a sawmill. Same as holding boom, receiving boom.

STORAGE DAM   A temporary dam to impound water for the spring drive.

STORE-BOUGHT   Not homemade.

STOVE   A log or lumber raft striking rocks and breaking up.

STOVELIDS   Pancakes or griddle cakes. Flats, flap jacks, washers, sweat pads.

STRAP IRON   Flat pieces of iron of various dimensions used by blacksmiths.

STRAWBERRIES   Beans.

STRAW BOSS   A boss in a logging camp. Same as head push, side push, working foreman.

STRAW KNIFE   A very heavy knife with two handles used for cutting straw bales or straw stacks in winter to use for bedding for oxen or horses.

STRAW PUSH   The man who took the job of the regular woods boss when he was away. Same as straw boss.

STRAY   An unmarked log which had floated away from the river drive.

STRAY LOGS   Logs that had been misdirected at the sorting works.

STRAYS   Same as stray logs.

STREAM JAM   A log jam on an island or rock in the center of a stream. Same as center jam.

STRIKING MAUL   A heavy wooden maul used in making homemade cedar shakes.

STRINGERS   Rafts of logs.

STRING OF FLATS   1. Griddle cakes, pancakes, stovelids, washers. 2. Railroad flatcars.

STRING OF LOGS   A section of a log raft usually sixteen feet wide, four hundred feet long. Rafts were made up of ten or more such strings.

—S—

STRIP That particular part of a forest assigned to a certain cutting crew.

STRIPS In early lumber grading, defective boards and strips were classified as "culls"; strips being very narrow boards.

STRUGGLE A dance.

STULL A timber used in a mine to support the sides and the roots of the passages.

STULLS Vertical support mine timbers approximately sixteen inches or more in diameter.

STUMPAGE The value of timber as it stood uncut in the woods. The value of the timber without the land.

STUMPAGE INSPECTOR Man who sees that the purchaser of standing timber cuts the right trees.

STUMP CUSHION Pancake.

STUMP DETECTIVE A forester who measured the waste in stumps and tops of trees.

STUMP LOGGING A logging operation in which the logs went directly from the woods to the mill, not stored in a pile. Same as hot logging.

STUMP PULLER Any of several types of mechanical devices used to uproot or remove stumps from the soil.

STUMP SET A convenient stump used as a base (vise) to hold a saw for filing.

STUMPSTEAD A pioneer home-farm in the "cutover."

STUMPWOOD Pulpwood bucked into small piles near the stump where cut.

SUCCOR An uprooted tree that has fallen into the stream. Same as sucker sweep, sweeper.

SUCKER 1. A limb growing out of a main trunk, forming a second small trunk of the tree. 2. A small log floating alongside a larger log into which a driver put his peavey to help balance while riding on the larger log. 3. A cleanup man who piled up the branches that were cut off felled trees. 4. A new jack in camp.

SUCKER LINE The rope that tied lumber cribs together on a river drive.

SUCKER SWEEP An uprooted tree that has fallen into the stream. Same as sweeper, succor.

SUGAR BUSH A stand of sugar maple (hard maple) from which sap is drained to make maple syrup and maple candy.

SULKY A pair of wheels, usually ten to fourteen feet high, used for transporting logs. Same as big wheels, timber wheels, katydid, logging wheels, high wheels.

SUN KINKS Appeared in skeleton tracks that had been laid in cold weather. Warm weather in spring could cause the rails to warp and make for derailments.

SUN TOAST Bread.

SUNDAY BOIL Same as boiling up.

SUPPLY HOUSE Also called wanigan on a river drive.

SURRY-PARKER LOADER Similar to the McGiffert loader.

SWABBER A man who greased skids to lessen friction of logs dragged over them.

SWAGE 1. Any of many heavy metal tools blacksmiths used to forge the shape of metal. 2. Tool used to flare or spread saw teeth on a crosscut saw.

SWAGE BAR Blacksmith's tool for shaping hot metal by holding the swage and striking with a hammer.

SWAGE SET Any saw fitted so that the teeth flare out laterally to make the cut wider than the back of the blade, thus preventing a binding.

SWALLOW TAIL STUMP   A stump with uneven surface because of poor notching. Generally cut by an inexperienced logger.

SWAMP   To clear the ground of underbrush, fallen trees, and other obstructions preparatory to constructing a logging road, a gutter road, or landing.

SWAMP ANGEL   A form of plow for cutting ruts in a logging road for the runners of a sleigh. The ruts were sprinkled with water and frozen to make an ice road. Same as rutter, gouger, groover.

SWAMP AUGER   A mythical tool. It was a common joke among the woodsmen to send a new hand back to the office to get a swamp auger. Often when the victim found out he was the object of a joke, he packed up his things and left camp. When the foreman wanted to fire a man, he told him to go into town and get a swamp auger.

SWAMPER   Man who cuts the limbs off a tree after felling, or cuts out a skidding road. Same as gutter man, trimmer.

SWAMPER'S AXE   A rather light, short-handled, two-bitted axe used by swampers to clear roads and skidways.

SWAMP FIRE BUCKET   A galvanized bucket with a stout wire handle used for fire fighting. Cone shaped so it could not be used for any other purpose; it was filled with water and hung up in all camp buildings for fire protection.

SWAMP GABBON   An imaginary animal to which snowshoe tracks were attributed.

SWAMP GRANTS   Public lands sold to individuals in any amounts.

SWAMP GRASS   Grass growing along streams in old beaver meadows and used for chinking between logs of the early camp buildings.

SWAMP HOOK   The hook with a sharp point at the end of the loading line. Used when rolling logs onto a travois or in loosening logs mired or frozen in the swamp.

SWAMP MOSS   Moss (probably sphagnum) used as chinking between the logs in camp buildings.

SWAMP OUT   Clean up or sweep out a building.

SWAMP RAT   Farmers who worked in the woods in the winter.

SWAMP SAUGER   A mythical, swamp-dwelling animal of great strength. When a log stuck in the mud or in a swamp, the men said it would take a swamp sauger to get it out.

SWAMP SHOE   A large wood or metal shoe fastened with clips to a horse's hoof to permit the animal to walk in swamps and over soft earth. Same as bog shoe.

SWAMP WAGON   A wagon whose wheels were cut from a section of a log and held together with iron bands or hoops, like tires on wheels. Used for hauling timber out of swamps.

SWAMP WATER   Tea. In early camps more tea was used than coffee.

SWAY BAR   1. Either of the two timbers used in coupling the front and rear sleds of a logging sleigh. The skids rested on the bar in loading logs. 2. The bar used to couple two logging cars.

SWEAT PADS   1. Pancakes. 2. Cotton-filled pads placed beneath a horse's collar to prevent chafing.

SWEDE HOOKS   Long-handled log tongs used for carrying heavy logs or timbers.

SWEDE SAW   A short saw with bow-type handle, generally used to cut pulpwood. Same as bow saw, Finn saw.

SWEDISH BRAIN FOOD   Snoose. Damp snuff for chewing. Same as Swedish conditioner powder, Copenhagen, Scandihoovian dynamite.

SWEDISH CONDITIONER POWDER   Snoose or snuff.

SWEDISH FIDDLE   A crosscut saw.

SWEENIED   A horse whose shoulder muscles have shrunk.

SWEEP   1. The natural bend in a log, generally applied to long, gentle bends. 2. A tree branch hanging low across a road or stream. 3. The curved bole of a tree. 4. Same as steering sweep.

— S —

SWEEPER   An uprooted tree that has fallen into the stream. It could sweep men off a raft or log; it frequently caused a jam. Same as a sucker sweep, succor.

SWEEPS   Long oars used on a bateau or lumber raft.

SWEETEN THE POOR BOX   Put some tobacco in the box reserved for jacks who had no money to buy their own tobacco.

SWELL BUTTED   A tree greatly enlarged at the base. Same as bottle butted, churn butted.

SWELL KNOB   Refers to the shape of the butt of an axe handle.

SWIFTER   A chain brought around two objects and twisted in the center to bring the two objects together.

SWINDLE STICK   A long hardwood ruler used in estimating lumber in logs or timber. Same as log scale rule, cheat stick, money maker, robber's cane, thief stick, scale rule.

SWING BOOM   A boom across the river. It could be closed to catch logs and opened to let trash through. Same as gate boom.

SWING BOOM LOADER   Same as swing jammer.

SWING BOOM JAMMER   Same as swing jammer.

SWING CANTED   Same as gunned.

SWING CHAIN   A short, stout chain used to attach logs to the undercarriage or axle of big wheels.

SWING DINGLE   A single sled with wood-shod runners and tongue with lateral play, used in hauling logs down steep slopes on bare ground. Same as loose-tongued sloop. Also, a sled used to haul lunch to lumberjacks in the woods.

SWING DOG GILLIARY or SWING DOG   A cant hook. The forerunner of the peavey. A large hook fastened onto the handle with a chain link which made the hook swing free. Unlike the peavey, the hook had to be set by hand. It was later called a gilliary. Reportedly invented by a man named Gilliary or Gillery who lived near Lilly, Wisconsin.

SWING DONKEY   A donkey engine used to supplement a yarder over a long haul.

SWING JAMMER   A log-loader with a swinging boom.

SWINGING BOOM GIN POLE   A gin pole having a boom which could be moved in a half arc for loading sleighs and railroad cars.

SWINGING BITCH   Jack's term for the cant dog.

SWINGING BUNK   A bunk on a sleigh held to beam by a king bolt or king pin.

SWING TEAM   In a logging team of six, the pair between the leaders and butt or wheel team.

SWITCH HOG   A small locomotive used on short-run log haul. A switch engine.

SWITCHING THEM IN   Keeping the logs moving with the current on a log drive.

SWIVEL HOOK   A round hook attached to a swivel.

SWIVELED TIMBER CARRIER   A pair of tongs attached to a bar to carry small logs. Lug hooks, come along.

–T–

Tractors

A lumberjack recalls, "The first winter a tractor was brought into the woods to help out the horses, the other old-time lumberjacks would not speak to us. Someone told them that the tractor was going to put them all out of work—that the tractor was going to take over and they would be going home. I don't know how they figured the logs would get sawed. Boy, we weren't very popular.

"It wasn't very long before the jacks saw we had a lot of power. The water tank would get stuck and we would unhook from our load and pull it out. Or maybe a load of logs would get alongside of the road and get stuck and we would pull them out. It wasn't very long before they got so dependent on the tractor I thought we would have to skid the logs. By spring we were the most popular gang in the woods. We were the 'big wheels'—they called us 'mister'. When a lumberjack calls you 'mister', you're pretty big."

TAFFLE or **TAFFLER**   A cookee, a cook's helper.

TAIL DOWN   Men who rolled logs to the loader, who loaded them on a sleigh or car.

TAILER IN   A cant hook man who rolls logs to the base of logs already in a deck.

TAIL HOOK   A skidding tongs. Same as dogs.

TAILING DOWN   Rolling logs down a skidway from the rear of the log pile. Rolling logs to the deck pile.

TAILING UP   Collecting derelict logs left stranded at the close of a river drive.

TAIL OF A SAWMILL   Refers to the sorting rack in a sawmill.

TAIL RACE   The water that has been diverted to a mill from a river or other stream. The water used to power the sawmill.

TAIL RIGGER   One who handled the carriage at a sawmill.

TAIL SAWYER   One who pulled slabs, boards, and cants from the head saw in a mill. Also cut unsalable parts out of lumber.

TAIL SHAFT   Propellor and drive shaft assembly of log tugs.

TAIL TREE   The No. 2 spar of a skyline hook-up.

TAKE TO THE WOODS or TAKE TO THE TALL TIMBER   1. A fast exit; to leave camp in a hurry. 2. To take off for camp after a spell in town.

TAKE 'EM BY THE EARS   In river driving, to slide a log to the river by the efforts of several men using peavies.

TAKE UP A SECTION LINE   A land looker's expression: to find and then follow a section line to a section corner.

TAKING A DRINK STANDING UP   He fell in the water.

TALLOW POT   A locomotive fireman.

TALL STUMPS   Left when trees were cut when there was deep snow on the ground, or cut wastefully by careless loggers.

TALLY   Same as scale

TALLY BOARD   A thin, smooth board used by a scaler to record the number or volume of logs.

TALLY BOY   Same as tally man.

—T—

TALLY MAN   One who recorded or tallied the measurements of logs as they were called by the scaler.

TANBARK   **1.** Any bark used in tanning. In the lake states it was generally the bark from the hemlock tree. The hemlock was sometimes called the tanbark tree. **2.** A compound derived from hemlock bark and used in tanning leather.

TAN-BARK WHISKEY   Cheap, raw whiskey

TANGLEFOOT   A cheap whiskey.

TANK   A large watertight box built on a sleigh. It sprinkled the logging road with water to make an ice road. Same as ice box, icer, ice tank, sprinkler, water box, water tank, water wagon.

TANK CONDUCTOR   One who had charge of the crew which operated the water tank or sprinkler, and who regulated the flow of water in icing logging roads.

TANK ENGINE   A locomotive with a water tank (called a "saddle tank") placed over the drivers for traction weight.

TANK HEATER   A sheet iron cylinder extending through a water tank or sprinkler in which a fire was kept to prevent the water in the tank from freezing while icing logging roads in extremely cold weather.

TANKING   The act of hauling water in the tank to ice a logging road.

TANK STOVE   Same as water tank stove or tank heater.

TANK TEAMSTER   Teamster who drove the teams on a water sprinkler.

TAG SLED   A short sled trailed behind the tote sled loaded with supplies.

TAPER OFF   To cut down on the liquor intake.

TAR PAPER ROOFS   In later day camps most of the buildings had heavy, black tar paper laid over heap boards as roofing—cheap and easy to apply.

T-BAR   Roll bar with sled tongue attached.

TEA KETTLE OUTFIT   A small mill crew.

TEAM   Logging parlance that expressed not only a set of 4–6 oxen that drew logs, but also the 3 to 4 or 7–8 men who kept one team employed.

TEAMSTER   A man who drove a team in a logging operation. Same as hair pounder.

TELESCOPE SUITCASE or **TELESCOPE BAG**   1. A fiber suitcase used by many jacks instead of a turkey, haversack, or bindle. Its cover could be lifted to permit much gear to be stowed. 2. An adjustable traveling bag made in two sections, the larger top section slipping over the other. Straps hold them together.

TENDER'S SHACK   A very small building at a dam where the dam watchman lived.

TENDING OUT   Rivermen stationed at places along the stream where a jam may form, to keep the logs moving.

TEN HOURS OR NO SAWDUST   A slogan used by striking sawmill workers.

TEN TON HOLT   A caterpillar tractor used in later-day logging operations made by the Holt Manufacturing Company.

THE CUTOVER   The area where forests once stood. Land where the timber had been removed.

THE JAM HAULED   The jam broke!

THE TURN OF THE NIGHT   Midnight.

THE WHOLE SHEBANG   Everything! As: "The whole shebang went up in flames!"

THIEF CAR   An empty railroad box car or flat car—There was no profit in pulling empty cars!

—T—

THIEF STICK   A long hardwood ruler used in estimating lumber in logs or timber. Same as money maker, robber's cane, cheat stick, scale rule, log scale rule, swindle stick.

THOUSAND   A thousand board feet is the unit of measure of logs or lumber.

THREE-LINK IRON BOOM CHAIN   The famous boom chain that replaced Manila rope in making up log rafts.

THREE TRIP ROAD   Three round trips per day.

THRIBBLE PINE   Same as buckwheat pine.

THROTTLE ARTIST   Steam locomotive engineer.

THROW   To topple over with wedges a tree that is being felled. Same as wedge a tree.

THROW A SAG IN HER   Set the cant hook firmly in the log.

THROWING THE SIEVE   To thin out the logging crew.

THROW LINE   A light rope attached to a dog hook, used to free the latter when employed in breaking a jam, a skidway, or a load. Same as trip line.

THROW OUT   1. A timber placed at the mouth of a slide to direct the logs. 2. The junction of railroad tracks in railroad logging. Same as frog.

TIE BEAM   A timber across the sluiceway of a dam to hold the side walls riding at top.

TIE BLOCK   Timbers that held two sleigh runners together in a bob upon which the bunks were laid.

TIE BUCKER   A man hired to load railroad ties on a car.

TIE CAMP   A camp housing jacks who only cut railroad ties.

TIE CUT   A log cut for use as a railroad tie.

TIE LOOSE   To untie a raft.

TIE-MAKER CAMPS   (Tie camps) Small camps, usually jobber camps used by cutters making railroad ties.

TIE-MAKER'S WEDGE   A wedge with hollow-ground sides.

TIE-UP LOADER   A manually operated log loader attached to a logging sleigh as a base; very similar to an A-frame jammer.

TIE WHACKER or TIE MAKER   A man who cuts ties in the woods.

TIE YARD   An area where railroad ties were piled ready for shipment.

TILLER END OF A PITSAW   End of the saw held by the top sawyer in a pitsaw operation.

TIM-BER   Traditional cry of warning when a tree was about to fall after being cut. Same as down the hill.

TIMBER   A man who cries "timber" in a saloon is announcing that he will buy drinks for the house.

TIMBER BEAST   A lumberjack or logger.

TIMBER BERTHS   A licensed cutting area.

TIMBER BIND   A saw cut tightening on a saw blade.

TIMBER BOUND   A faller with a stuck saw.

TIMBER CARRIER   A pair of tongs attached to a bar used by two men to carry small logs. Same as lug hooks, timber grapple, timber hook.

TIMBER CONTRACT   The right to purchase standing timber without getting title to the land. Same as timber right.

TIMBER CRUISER   Same as cruiser.

TIMBER DOG   Same as hewing dog.

TIMBER FITTER   1. A man whose job was to size up a tree, determine the exact spot where it was to fall, and notch it accordingly. 2. Boss of a falling crew or cutting crew.

TIMBER GRAPPLE   A pair of tongs attached to a bar used by two men to carry small logs. Same as lug hooks, timber carrier, timber hook.

–T–

TIMBER HOOK   A pair of tongs attached to a bar used by two men to carry small logs. Same as lug hooks, timber carrier, timber grapple.

TIMBER LIMITS   A licensed cutting area.

TIMBER LOOKING   Estimating board feet of timber in a given area prior to buying and later cutting.

TIMBER PIGS   Big operator lumbermen.

TIMBER PIRATES   Those who stole logs as they drifted downstream, were piled on landings, or were grounded during a river drive.

TIMBER RIGHT   The right to purchase standing timber without getting title to the land. Same as timber contract.

TIMBER SCRIBE   Stylus used by a surveyor to mark a witness tree.

TIMBER SHAVE   A draw shave with curved blade for removing bark and wood from round logs.

TIMBER SNAKES   Crawler type steam log haulers such as the Lombard and Phoenix.

TIMBER STAMP   Same as log-marking hammer or stamp hammer.

TIMBER THIEVES   Same as timber pigs.

TIMBER TONGS   A pair of hooks attached to a handle used to move heavy pieces such as bridge timbers.

TIMBER TRESPASS   Cutting timber on someone else's land. Stealing timber.

TIMBER WHEELS   A pair of wheels, usually ten to fourteen feet high, used for transporting logs. Same as big wheels, katydid, logging wheels, sulky.

TIMBER WOLF   Jack's affectionate name for the timber cruiser.

TIME AND TALLIES   Camp records kept by the camp clerk.

TIME BOY   An office boy who gathered time and purchase slips from the company's mill, camp and yard foremen, and brought them to the central office for payroll and account bookkeeping.

TIME CHECK or **TIME CHECKS**   1. A kind of script that was issued to loggers which could be discounted at banks or business establishments. 2. A paper "promise to pay" given jacks in spring in lieu of cash, redeemable in cash in the fall when logs were sold. Local stores took off 10% for cashing these time checks in the summer.

TIME SLIPS   Pay checks.

TINKER   A wood butcher or carpenter.

TINKER SHOP   A small log building in some logging camps where pots and pans were mended and axes and saws were repaired and sharpened.

TIN PANTS   Heavy, water repellent trousers. Denim trousers.

TIRE BENDER   A mechanical device used by wheelwrights to affix the curve in wagon wheel tires or rims.

TISSUE FLIPPER   A railroad conductor.

TIT   Usually, the hand throttle on a donkey or a logging locomotive.

TOBACCO CUTTER   A hand-operated machine, operated like a small guillotine, that cut off plug tobacco in chunks by the inch.

TOBACCY   Jack's term for tobacco.

TOE PILING   Sharpened poles or timbers which are driven next to the upstream face of the mudsills of a dam to prevent water from getting under the foundations.

TOE RING   The heavy ring or ferrule on the end of a cant hook. It had a lip on the lower edge to prevent slipping when a log was grasped.

TOGGLE   Metal bar with a hole in the middle by which it was attached to a line or chain. If the toggle was turned lengthwise, a chain could pass through and then hold when turned horizontally.

TOGGLE CHAIN   A short chain with a ring at one end and a toggle hook and ring at the other, fastened to the sway bar or bunk of a logging sled, and used to regulate the length of a binding chain. Same as bunk chain.

-T-

TOGGLE HOOK   A hook with a long shank, used on a toggle chain.

TOGGLE KNOCKER   The man who knocked logs loose from the binding chain used with big wheels.

TOGGLE WITH RING   A boom chain having a toggle at one end and a ring at the other.

TOMBSTONE   A stump with part of the tree left due to the tree's splitting when falling. Same as barber chair.

TONGING   Handling logs with skidding tongs.

TONGS   Large sharp hooks on the end of a chain or cable for gripping logs.

TONGS TABLE   A yard-square, wood table which held the blacksmith's tongs.

TONGUE HOOK   A metal hook attached at the lead end of the reach or tongue of a logging sleigh.

TON TIMBER   A ton or load was twelve inches square and forty feet long. It was generally hewn timber.

TOOK WATER   The camp bully who "bit off more than he could chew" and then chose to quit fighting was said to have took water.

TOOL BRAND or BRANDING IRON   A metal iron mounted on a rather long handle with embossed ownership initials or names used when heated to identify company tools.

TOOT   As "he's off on a toot." He's off on a spree.

TOOTHPICK WOOD   Wood from which toothpicks are made—usually white birch.

TOP BRASS   Owner or manager of any logging operation.

TOP CHAINS   Chains used to secure the upper tiers of a load of logs after the capacity of the regular binding chains had been filled.

TOP DECKER   Same as top loader.

TOP LOAD A load of logs piled more than one tier high, as distinguished from a bunk load.

TOP LOADER or TOP DECKER That member of a loading crew who stood on the top of a load and placed logs as they were sent up. Same as sky bird, sky hooker.

TOP LOADER'S CANT HOOK A cant hook having a heavy, flat pry point at the tip instead of the usual metal heel.

TOP LOADER'S SEAT A small seat that could be attached to a jammer boom to lift a top loader down from the top of the log pile on a sleigh.

TOP LOADING 1. The work of arranging the logs on top of the car or sleigh. Same as sky hooking. 2. Shingles, laths, pickets, etc. placed on top of lumber in transit on rafts.

TOP MAN Foreman, boss, top brass, main say.

TOPOUT A SLED Place the final tiers of logs on a partially loaded sleigh.

TOP SAWYER Sawyer standing on the platform in a pit sawing operation.

TOQUE A cap or hat worn by some jacks.

TORCH Dry, pitch pine shavings fashioned into a torch to be used by the river pigs at night.

TOTE To haul supplies to a logging camp.

TOTE ROAD A supply road to the camp. Same as pike, portage road.

TOTE SLED A sled shod with wood, used for hauling supplies over bare ground into a logging camp. Same as jumper.

TOTE TEAM Horses used to take supplies into camp.

TOTE TEAMSTER Driver who brought supplies into camp.

TOTE WAGON A very heavy wagon used to haul supplies to camp via the tote roads.

—T—

TOTING   Hauling supplies to camp.

TOW CHAIN   A long chain attached to oxen or horses and to the ends of logs in skidding.

TOW HEAD   A river term meaning a little sand bar or an island with grass and willows growing on it. If a raft struck the bar, it doubled like saddlebags around the bar.

TOW HILL   A steep hill which required an extra team of horses to help pull up a loaded sleigh.

TOWN CLOWN   A small town policeman.

TOWNIES   Local town toughs.

TOW OF LOGS   A number of logs being skidded by horses at the same time.

TOW TEAM   An extra team stationed at an incline in a logging road to assist the regular teams in pull the loaded sleigh up the hill. Same as snatch team.

TRACE   That part of the horse harness that connects the hames with the whiffletree. Same as trace chain.

TRACE CHAIN   A short chain from the traces to the singletree or doubletree when animals were used in woods operations. Or, simply a short piece of chain.

TRACK DRILL   A mechanical, hand powered drilling machine designed to cut holes in railroad track side in order to attach the slide plate in joining two pieces of track.

TRACK GAUGE   A metal or wood gauge so calibrated as to measure the angles in a curve of railroad track.

TRACK PICKER   A crane-like device mounted on a railroad flat car used to pick up the rails on an abandoned railroad track.

TRACK WALKER   Man hired to patrol the railroad tracks looking for broken rails, washouts, and any other dangerous road conditions.

TRACTOR LOGGING A logging operation where skidding is done with crawler tractor power.

TRADING STOCK Same as squaw wood.

TRAILERS Several logging sleighs hitched one behind another and pulled by four or five teams of horses driven by one man.

TRAILER SLEIGH Same as trailers.

TRAILING DOWN Bringing logs from piles in the woods.

TRAM, TRAMWAY or TRAM ROAD A light or temporary railroad for the transportation of logs, often with wooden rails and operated by horsepower. Wheels were concave.

TRAMP LOGGER One who changed from one job to another.

TRAMP THE ROAD To pack down the snow by tramping with men or oxen so that the road would freeze. Sometimes ten or fifteen men would be put to work to tramp a road after a big snow.

TRANES A Canadian term for sleighs.

TRANSFER The sorting works where first lumber grading is done.

TRAVELING DANDRUFF Lice.

TRAVELLER A metal wheel with a handle used to measure the distance around the felloes of wagon wheels.

TRAVELLING BLOCK A pulley designed to move along a cable while supporting a load.

TRAVIS POLE A pole laid crosswise to connect two floating logs. Water-logged logs were wired to this pole to float them to the mill. It was also used to transport deadheads.

TRAVOIS or TRAVOY Sometimes incorrectly spelled travoy. A dray used to haul logs from woods to skidway. Generally made from the natural fork of a tree with a crosspiece bolted midway in the V. Also made from a birch or maple tree with a large root turning at the proper curve for the front. In use, one end of the log rested on the travois and the other end dragged on the ground. The same as crazy drag, go-devil, snow snake, jinnie.

TRAVOIS DRAY   Same as travois.

TRAVOIS ROAD   A long skid road used to drag logs from woods to skidway. Same as dray road, drag road, runway.

TREE SQUEAK   An imaginary bird to which the noise made by trees rubbing together was attributed.

TRIM BLOCK   Waste at a sawmill composed mostly of trimmings.

TRIMMER   1. One who cuts the limbs off fallen trees. Same as swamper, gutter man. 2. A millsaw that trimmed lumber into standard lengths. 3. Operator of trimmer saws.

TRIMMER TAILERS   Workers who handled trimmings at a trimmer saw.

TRIMMINGS   Wood debris from edgings and slabs at a sawmill, or waste left in the woods when trees were felled and tops and limbs were removed before bucking. See slash.

TRIP   1. To topple over with wedges a tree that is being felled. Same as wedge a tree. 2. A single trip and return made by one team in hauling logs. Same as turn.

TRIP A DAM   To remove the plank that closes a splash dam.

TRIP BOOM   1. A log placed across a stream to hold the logs in a drive until the crew was ready to start them down. 2. A boom strung a safe distance above a dam, falls, or fast rapids that could be opened to let logs through when the stream was clear, or closed quickly in case a jam formed below it.

TRIP HER   An expression used when quitting the job.

TRIP HOOK   1. A hook attached to a chain which was wrapped around a log being loaded by a jammer. The trip hook released the chain when the log was in place. Same as slip hook. 2. A hook with a fixed handle used with a skidding harness to drop the skidding chain merely by releasing the trip hook by a kick with the boot.

TRIP LINE   A light rope attached to a dog hook, used to free the latter when employed in breaking a jam, a skidway, or a load. Same as throw line.

TRIP SILL   A timber placed across the bottom of the sluiceway in a splash dam, against which the plank rested to close the dam.

TRIP STAKE   Stakes fitted into rockers of a logging sleigh which could be removed to dump logs at a landing.

TROUGH ROAD   Narrow-gauged ice roads where horses travel in iced ruts with narrow gauge sleds.

TROUGH ROOF   A roof on a logging camp or barn, made of small logs split lengthwise, hollowed into troughs and laid from ridge pole to eaves. The joints of the lower tier were covered by inverted troughs.

TROUGHED-OUT LOGS   Logs partially hollowed out like a trough prior to making a scoop roof. See trough roof.

TRUCK   A heavy wagon used to haul logs, either with animal power or power traction.

TRUCKSTERS   Sawmill workers who piled lumber onto lumber buggies. Same as buggy loaders.

TURKEY   Packsack or any kind of sack such as a meal sack in which a lumberjack carried his belongings. A duffle bag, kennebecker, tussock, keister.

TURN   1. A unit of logs being yarded. 2. Several thousand feet of logs held together by dogs, to be hauled along a skidway. 3. A single trip and return made by one team in hauling logs; e.g., a four-turn road is a road the length of which will permit only four round trips per day. Same as trip.

TURN AROUND   A cleared area in which a sleigh, truck, wagon, tractor, or other skidding device could be turned around at the end of a road.

TURNIP   A watch, generally a large watch tied on pants with a leather thong.

238

—T—

TURN OUT   A short side road from a logging sleigh road to allow loads to pass.

TURN OUT THE CREW or **TURN OUT THE MEN**   To get the lumberjacks out in the morning.

TURN TABLE   A section of twin rails long enough to support a logging locomotive acting as a platform which could be turned in order to rotate a locomotive.

TUSSOCK   A packsack. Same as duffle bag, kennebecker, keister.

TWIN NECK YOKE   A short neck yoke that goes on each end of the long neck yoke to assist in holding back a load. Same as whiffletree neck yoke.

TWIN SLEDS   Sleighs for hauling logs. Same as two sleds, wagon sleds, sled.

TWIST   1. Same as twitch. 2. A very early brand of chewing tobacco popular with jacks.

TWISTER   A device for moving heavy objects in logging. It consists of a rope or chain within a turn of which a lever is inserted and twisted. Same as Spanish windlass, a turnbuckle.

TWITCH   1. To skid logs or full-length trees to yarding area. Same as snake. 2. A short chain. 3. A loop of rope attached to a short, hardwood handle. When the loop is placed over the nose of a balky horse and the handle twisted it causes pain in the horse and he will "behave" while being shoed.

TWITCHER   Same as twitch.

TWITCHING CHAIN   Short chain having round hook on one end and grab hook on the other used in skidding logs.

TWITCH ROAD   A narrow road or trail for twitching logs.

TWO-BUNK DRAY   A dray with double bunks used to haul railroad ties, posts, etc.

TWO HANDER   A large stick used with both hands to pound horses to make them pull harder or to obey orders. Same as brolo.

TWO-HORSE YARDING ROAD   A yarding road where two horse hitches are used in the skidding operation.

200 FOOT TO THE LOG   The logger's optimum number of board feet to the log.

TWO-SLED, TWIN SLED   Refers to a logging sled consisting of two bobs chained together in tandem.

TWO SLEDS   Sleighs for hauling logs. Same as sled, twin sleds, wagon sleds.

TWO STREAKS OF RUST   A logging railroad.

TWO TRIP ROAD   Number of round trips per day a logging team hauling logs could make. A two trip road meant that the distance one way was about six miles.

Eats

Lumberjacks had what they claimed were the best eats in the world. For breakfast there was fried potatoes, prunes, bacon or sow belly, pancakes, fresh biscuits, oatmeal, pie left over from the day before, and doughnuts. Always doughnuts. And then there was tea. Not much coffee was served in the early pine logging days.

There was no coffee break for the lumberjack. At eleven-thirty the gabriel was blown. The men came into camp, if it was close, but generally the shanty boss or cookee hauled the dinner to the woods in boxes on a jumper. Dinner was the big meal. There was red horse or roast pork, murphys or rutabagas, bean hole beans, dried apricots, peaches, currants, raisins or prunes, fresh-baked raisin or pregnant woman pie, cookies, doughnuts, coffee cake or raisin cake, mountains of bread and jam, and java or tea.

For supper they let up a little. There was soup, hot biscuits or johnny-cake, cold meat or hash or beef stew, potatoes, bean hole beans, rice pudding, pie or pound cake, blackstrap or brown sugar cookies, and leftovers from dinner.

UNCLE   A superintendent.

UNDERBRUSH   Any kind of plants or bushes growing beneath the trees in a forest.

UNDERCUT   The first cut in falling a tree. It determined the direction of the fall. Same as nick, notch.

UNDER CUTTER   A skilled woodsman who chopped the undercut in trees so that they would fall in the proper direction. He used an axe lighter than standard weight.

UNDERSHOT WHEEL   Refers to a water sawmill with power being generated by a water wheel turned by water flowing under it, not over it as in the case of the overshot wheel. See overshot wheel.

UNION DRIVE   A drive of logs which belonged to several owners, who prorated the expenses.

UNION LEADER   A popular brand of camp tobacco.

UNLOAD   Jump, getting off from a sluiced road. To jump off a runaway to get out of danger.

UNRAVEL   To clear jammed pulpwood from a stream in a drive.

U.P.   Upper peninsula of Michigan.

UP AND DOWN SAW   Upright saw used in water power sawmill. Same as "up today and down tomorrow." Came from Russia.

UP A TREE   Stuck, treed, hung up on any problem.

UPHILL CLEVIS   A sleigh brake used to keep a loaded logging sled from slipping backwards on a hill.

UPPERS   Number one grade of white pine lumber.

UPRIGHT ROLLER   The cross bar of a logging sleigh into which the tongue was set. Same as roll, roller.

UPRIGHT SAW   A sash saw or frame saw.

UPRIVER   The cutting areas far up on the stretches of the rivers.

242

-U-

UP THE POLE   A logger on the water wagon, that is, one who did not drink.

UP TODAY AND DOWN TOMORROW   An upright saw used in water power sawmill. Same as up and down saw.

USING THE SPOON   A term applied when a teamster gave medicine to his horse.

On the Town

After the spring breakup when the logs had been driven down the river to the mill, the lumberjack usually went on a spree. And a great spree it was. The lumberjack was interested in only two things, those which came in a bottle or in a corset. He had also a set of standards of his own, however, and it was his boast that he never offended a respectable woman. Fights were common; they were a form of entertainment. Towns were rough as a result of the lumberjack's desires and demands. An oft-repeated statement in Wisconsin was "Hayward, Hurley, and Hell," in the order named.

In 1870 there reportedly were 306 saloons and/or sporting houses in Saginaw, Michigan. In Bay City, Michigan, Water Street was so notorious it was called Hell's Half Acres.

VALUE To estimate the amount a value of standing timber. Same as estimate, cruise.

VALUER One who cruised timber to estimate its value. Same as cruiser, estimator, land looker.

VAN 1. The small store in a logging camp in which clothing, tobacco, and medicines were kept to supply the crew. Same as wanigan. 2. A large iron chest banded with metal strips for reinforcement used to store clothing, tobacco, and medicines in the early shanty days.

—V—

VAN BOOKS   Camp store record books.

VEIN OF PINE   A grove or clump of pine growing within the deciduous forest. See pockets of pine.

VELOCIPEDE   Same as pede.

VENEER   A thin layer of wood made by turning a log in a lathe-like machine in a veneer mill.

VENEER LINE   A famous woods spur of the Mellen Lumber Co., running southeast out of the city of Mellen.

VIRGIN FOREST or VIRGIN TIMBER   The original stand of timber. An uncut stand of timber.

VOLUME   Amount of timber in a given area; in a tree, a log. Such volume is always expressed in board feet.

V PLOW   A plow made of two poles or planks fastened together at one end and kept apart at the other by a spreader. Used in breaking out and smoothing winter roads.

Clothing of the Lumberjack

Lumberjacks wore heavy two-piece red wool underwear, heavy Chippewa or Malone stag pants, two or more pairs of warm wool socks inside packs, one or two wool shirts, a heavy mackinaw, wool cap, and two pair of mitts that kept him warm in many-degrees-below-zero weather. Stag pants were worn over the packs, not inside them, to keep out the snow. The local shoemaker made the packs, which had leather uppers eight to sixteen inches high sewn to rubber bottoms. Some of the early lumberjacks wore all-leather, well-oiled boots.

WADE   One of the early drag saws.

WAGON BOX   Refers to a small tool box carried on the tote wagons for making simple repairs to wagons while traveling the tote roads.

WAGON SLED   A sleigh for hauling logs. Same as sled, twin sleds, two sleds.

WAGON WHEEL BRAKE   A shoe-like, iron device, sometimes called a "wheel shoe" used to slow down wagons on a downhill run.

WAILER   A jack who "wailed" when singing logger's ballads.

— W —

WALKER or WALKING BOSS 1. The superintendent of two or more logging camps. 2. A man elected from among the river drivers to be boss of all the crews on the drive. He was elected from members of different crews when they met at the junction of a river.

WALKIN BOSS Same as walking boss.

WALKING RUBBERS Packs with rubber feet and leather tops. Shoe packs.

WALKING THE COOK Getting rid of a very poor cook.

WAMPUS CAT An imaginary animal to which night noises were attributed.

WANE A board with a little bark left on the edge.

WANGAN BOX 1. The padlocked box serving as the camp store in the very early logging days. 2. A place where camp was made for the night on river drives.

WANIGAN or WAGON Different ways of spelling wanigan are: wanegan, wangan, wannegan, wanngan, wannigan, wanagan. Generally spelled wanigan in the lake states and wangan in the New England states. The origin is Indian. Wholly different meanings are: 1. Where the camp stores were kept. A place where lumberjacks could buy snuff, clothing, shoes, mittens, tobacco, underwear, jackets, pants, mackinaws, or other articles at cost. Same as van. 2. The payroll charges for such goods. 3. It was also the cook's raft which followed the log drive down river, and tied up to a tree on shore overnight. Same as shanty boat, ark. 4. Sometimes it made camp on a river drive and the men were paid. In eastern states, a wagon that carried food and supplies was called a Mary Anne.

WANIGAN MAN One who prepares camp for river drivers.

WARMING FIRE A small fire built in cold or wet weather by loggers to keep warm in the woods, especially at noon while eating lunch.

WARP To move logs with a headworks.

WARPING TUGS Steam tugs used in pulling log booms.

WAR SHOES   Surplus U.S. Army shoes sold after the close of the Civil War and often worn by jacks.

WASHED HIS CLOTHES   A river driver fell into the water.

WASHER   Doughnut. Same as sinker, stovelids, flats.

WASH-UP SINK   A bunkhouse dry sink where jacks could wash their hands and faces.

WATCH WILD   An early warning cry when timber was being cleared: "Stand Clear!"

WATER BOX   A watertight box built on a sleigh. It sprinkled logging roads with water to make an ice road. Same as sprinkler, ice box, icer, ice tank, tank, water tank, water wagon.

WATER CAR   A railroad car with water tanks mounted on it to supply the locomotive with water.

WATER CART   A sprinkler for watering ice roads.

WATER FLUME   A trough of wood made to divert water over a water wheel at a water sawmill.

WATER GATES   Adjustable barriers in dams to hold or release water.

WATER CREW   Those men in a driving crew who got the logs started downstream.

WATERING TROUGH   A pine log about 24 feet long hollowed out with an adze to hold water for horses and oxen to drink.

WATER LADDER   Two poles fastened together between which a barrel slides in filling a water tank or sprinkler. The barrel was pulled up the slide by a horse.

WATER MARK   A symbol stamped in the end or side of a log with a stamping hammer to identify ownership. Same as log mark, side mark.

WATER MARKS   Same as side marks.

WATER POWER SAWMILL   A sawmill where the saw was driven by water power.

WATER SCALE   Scale of logs in a boom or raft. Talley.

WATERSHED   Total area of land drained by a given stream on river.

WATER SLIDE   A V-shaped runway of lumber to bring logs out of mountains or steep hills. To transport logs. Same as flume, wet slide.

WATER TANK   A watertight box built on a sleigh. It sprinkled logging roads with water to make an ice road. Same as sprinkler, ice box, icer, tank, ice tank, water box, water wagon.

WATER WAGON   A sprinkler to ice the roads. Had a tongue at each end so that it could be pulled both ways. Same as ice box, icer, ice tank, sprinkler, tank, ice tank, water box, water wagon.

WATER WAGON   A sprinkler to ice the roads. Had a tongue at each end so that it could be pulled both ways. Same as ice box, icer, ice tank, sprinkler, tank, water box, water tank.

WATER YOKE   A wooden yoke to fit the shoulders of a man. Used in lumber camp when a man was carrying two pails of water or two boxes of grub.

WEDGE A TREE   To topple over with wedges a tree that was being felled. Same as trip.

WEDGES   Loggers always carried two wedge-shaped irons fastened together by a rope or wire. He also generally carried a wedge hammer which was a two- or three-pound maul.

WELDING HEAT   Heat sufficient to enable a blacksmith to form a weld in metal.

WELL AUGER   A very heavy, metal auger about 8–10 feet in length used to drill a camp well where the water table was high.

WENT OUT   Off to town on a drunk or a toot.

WET ASS   A dunking in cold water during a river drive.

WET SLIDE   A V-shaped runway of lumber to bring logs out of mountains or steep hills. Same as flume, water slide.

WET YOUR WHISTLE   Have a drink!

WHARFAGE   Refers to charges levied by towns along rivers charging a fee for rafts to tie up and dock temporarily at the river front of the towns.

WHEEL   In the pine days, to haul logs by means of big wheels.

WHEEL CAMP   Any camp using big wheels to transport logs.

WHEEL'ER   To quit or leave camp.

WHEELERS   The team closest to the logs in a log haul. Same as butt team, pole team.

WHEEL OF FORTUNE   A grindstone.

WHEEL ROAD   Any road over which a wagon could travel, usually a tote toad.

WHEN HEAD HIT   When the head of water arrived at the logs in the water to be driven.

WHEN HEAD RAN OFF   When water released by the dam was used up, the logs stopped and had to wait until more water built up in back of the dam.

WHIFFLETREE or WHIPPLETREE   The bar or singletree to which the horse's harness traces were fastened.

WHIFFLETREE NECK YOKE   A heavy logging neck yoke on end of pole or tongue to the ends of which short whiffletrees were attached by rings. From the end of the whiffletree, wide straps ran to the breeching, thus giving the team added power in holding back loads on steep slopes. Same as twin neck yoke.

WHIP LOADER   Same as side jammer.

WHIPSAW   1. An up and down saw with one man in a pit and the other on top of the log to saw it into lumber. 2. A bad situation, a deal that could get a man both coming and going.

WHIRLING THE ROCK   Turning the grindstone while sharpening axes.

WHISKERS   1. Jaggers on worn wire chokers or cable. 2. Fibers adhering to edges of shavings indicate that the rakers are cutting below the depth at which the teeth have scored the wood and that the rakes, therefore, should be shortened.

WHISKEY JACK   1. A hydraulic jack filled with alcohol as a nonfreezing fluid. 2. The Canadian jay bird, a camp robber.

WHISTLE PUNK   A workman who blew signals for yarding crew.

WHISTLE WATER MAN   Logger who was good on river drives.

WHITE HORSE   Waterfalls or fast rapids.

WHITE LEAD   Painted on cracked and swollen toes and chapped hands.

WHITE MEN   Some lumberjacks called themselves white men and would not bunk in the same room with Poles, Finns, Russians, Montenegrins, and other foreigners.

WHITE PINE   Pinus strobus.

WHITE PINE UPPERS   White pine boards graded as "selects"—very top grade.

WHITE WATER   The current in a stream fast enough to make white foam. Same as quick water, rapid water.

WHITE WATER MAN   A river driver who was expert in breaking jams on rapids or falls. One who could ride logs through fast water.

WHITTLING   Most jacks carried a pocket knife. Much of their idle time was spent whittling as a hobby. Such items as wood or birchbark snuff boxes, spruce gum boxes, cedar fans, and chains made from basswood or white pine.

WHOPPERS   Tall tales told on the deacon seat.

WIDOW MAKER   A tree lodged against another tree, or a hanging branch suspended in the top of a tree. A hanging branch could be released by wind or otherwise fall without warning, or spring back in the falling process. Many men were killed dislodging such trees. Same as sailer.

WIGWAM   In felling trees, to lodge several in such a way that they support each other.

WILD CENTER   A spot where a log would catch on a rock during a river drive, causing a jam.

WILDER   A brand of chewing and smoking tobacco. Same as Badger.

WINCHY or WINCHER   A winch operator.

WIND BREAKER   A mackinaw.

WINDBROKE   An animal strained by overwork and unable to breathe properly. Sometimes referred to a man with asthma.

WIND FALL   1. An area where the trees had been blown down by winds. 2. A single tree that had been blown down. Same as blow down, wind slash.

WIND FALL BUCKER   A man who was engaged in sawing up trees that had been blown down by wind. Fallers said wind fall buckers were loggers who not only talked to themselves, they also answered themselves.

WIND SHAKE   A crack in timber due to frost or wind. Same as shake.

WIND SLASH   A tree that had been blown down. Same as wind fall, blow down.

WINDTIMBER   Beans served any style in the camps.

WING   A small log jam intentionally caught on a rock or other obstruction for the effect of a sluiceway.

WING DAM   1. A pier built from the shore, usually slanting downstream, to narrow and deepen the channel, to guide logs past an obstruction, or to throw all the water on one side of an island. Same as pier dam. 2. A dam built from one river bank outward in a river to turn logs from the river into the mill race.

WING DING   The spree a lumberjack went on when he was stake bound.

WINGED LOGS   Logs piled out in the river and held in place to help form a channel through which a log drive could pass.

WING JAM   A jam formed against an obstacle in the stream and slanted upstream until the upper end rested solidly against one shore, with an open channel for the passage of logs on the opposite side.

WING LOG   An outside log of a load above the bunk log.

WINTER OF THE BLUE SNOW   During Paul Bunyan's logging operation, it was said that one winter the snowfall was blue.

WISCONSIN STICKER   A lumberjack who worked until he made his stake. He did not move to another camp very often.

WITCH   A chain and lever that bound the lumber together in a lumber raft.

WITCH-PLANKS   Top traverses or stringers of a log raft.

WITCH STICK   Twig or branch of alder brush used to locate water for the logging camp.

WITHE   A tough flexible twig, generally a willow, used to bind things together.

WITNESS TREE   A tree a surveyor marked with a stylus to locate the corner post. Placed on section and quarter-section corners. Same as bearing tree.

WOBBLIES   Same as wobbly.

WOBBLY   A member of the Industrial Workers of the World.

WOBBLY HORRORS   Fears of labor trouble by an employer.

WOLF TREE   A large tree overtopping and smothering the young growth. Often a limby, low-value specimen.

WONDER WORKER   A liniment used for everything from broken legs to toothaches.

WOOD   Common name for pulpwood.

WOOD BUCK   Man who sawed firewood for camp.

WOOD BURNER   1. A wood-fired locomotive or donkey. 2. Anything old fashioned.

WOOD BUTCHER or **WOOD CARPENTER**   1. The logging camp carpenter who made logging equipment out of wood. 2. Caustic name given to some of the early timber barons.

WOOD CHOPPER'S WEDGE   A wedge with a rather large angle.

WOODEN RAILROAD   A temporary wooden railroad using wooden, peeled poles as rails. Same as tramway.

WOOD HEAD   A lumberjack or logger.

WOOD HICK   A lumberjack.

WOOD HOOK   A pulp hook. Same as birch hook.

WOOD PECKER   A poor chopper. Same as beaver.

WOODS BOSS   Logging superintendent. Same as main say, bully, head rig, king pin, bull of the woods.

WOODS SPUR   Railroad tracks laid to a cutting operation or woods landing.

WOOD TICKS   Eight-legged Arthropods bothersome to jacks during spring cutting seasons.

WOOD UP   To load the wood box on a locomotive.

— W —

WORKING FOREMAN   The boss of a lumber operation. Generally one who worked in a small crew. Same as head push, side push, straw boss.

WORKING OFF   Freeing a lumber or log raft stuck on a sand or mud bar or on a rapids.

WORKS   The entire cutting area of a logging operation.

WORK TRAIN   A locomotive, cars and caboose, used in laying new track, taking up old rails on an abandoned track, or doing track repairs.

WRAPPER   A binding chain wrapped around the logs in the lower half of the load to hold them tightly to the sleigh.

WRAPPER CHAIN   A chain used to bind together a load of logs. Same as binder chain.

WRECKING FROGS   Devices used to get railroad engines and cars back on the tracks after derailments.

WRESTLE   A dance.

The Broad Arrow Policy

The use of the log mark denoting ownership is in this country older than the use of the brand on cattle. In early colonial days, Queen Anne's surveyor general marked with a "broad arrow" some of the finest pine and spruce in New England in an attempt to reserve them for spars and other uses for the royal navy. These efforts to appropriate American property were much resented by the hot-blooded colonists, and the "broad arrow policy," though unsuccessful, was one of the issues that brought on the Revolutionary War.

X-TREE In colonial days, any tree marked with an **X** was to be saved as a spar tree for the queen's navy and not taken by fallers. Similar to broad arrow mark.

–Y–

Drinking Songs

And now my song it is complete,
I think it is your turn to treat;
My favorite drink's hot whiskey sling,
but I will drink most anything!

Lumberjacks we are.
If you have timber that is up,
We'll cut it down.
If you have timber that is down,
We'll cut it up.
Lumberjacks we are;
Give us a drink!

I don't give a damn
For any damn man
Who don't give a damn for me.

YANKEE-TOE This was a tripping maneuver which lumberjacks scuffling in the bunkhouse used to get the opponent off his feet.

YANNIGAN A bag in which a lumberjack carried his clothes. A packsack.

YAPS Crazy, out of his mind.

YARD   The place where logs were assembled, generally before transporting them to the river. Same as landing.

YARDED WOOD   Pulpwood twitched or skidded to the yard or landing.

YARDER   An engine used in assembling logs.

YARDING   Moving logs to a central spot.

YARDING ROAD   Same as skidding road or skid road.

YARDING SLED   A short heavy sled used to haul logs to the yard or landing. One end of the log was placed on the bunk while the other end dragged on the ground. Same as dray.

YELLOW PINE   Norway pine.

YELLOW SOAP   Fels Naptha soap.

YOHN LINIMENT   One of the jack's many favorite liniments.

YOKE   Originally meant team of oxen. The wooden frame by which oxed were hitched together.

YOKE OF OXEN   A team of oxen.

YOU KNOW   A tin wash basin just for taking up collections. A penny was soldered to the bottom to remind the lumberjacks that they couldn't give less than a penny when it was passed.

YOUNG GROWTH   Young trees coming up after a logging operation.

The End of an Era

There will never be another sleigh haul and there will never be another log drive in the lake states, but this group of lumbermen will never be forgotten. It is all a part of history now, the story of the building of the great Midwest.

ZEKE   The name of Paul Bunyan's bookkeeper, who used a peavey handle as a pen holder.

ZIPPO   Corrupt spelling for gyppo. A logger who operated on a small scale. Same as chin-whiskered jobber.

NICKNAMES OF LUMBERJACKS

A camp "ink slinger" was having trouble keeping the camp records straight because there were three men working at the same camp with the same name. None had a middle name; the men did not work on the same day, but as each one arrived the problem grew.

After studying each man's appearance and checking starting dates, the ink slinger solved his problem simply by entering the following names in his book:

1. Dirty Joe Michaud
2. Bald Headed Joe Michaud
3. Joe Come-Lately Michaud

A great number of lumberjacks were known by nicknames. These names were often descriptive of their work or indicated certain characteristics of the individuals.

It was not uncommon for a woods worker to spend his entire lumberjack life known by a name not his own. Indeed, sometimes a lumberjack assumed a different name because he wanted to lose himself in the North Woods away from his family or the law. A man in a logging camp was taken at his face value, his past being his own business, and his alone. More often, nicknames were given to the lumberjacks by their fellow workers.

A few of these nicknames are preserved. All of these are actual names of lumberjacks who roamed the woods:

Angus the Pope	Bill the Moocher	Buckskin Pants
Battle Axe Nelson	Black Dan	Bug House Lynch
Billie the Bow	Blueberry Bob	Bullblock Andy
Bill the Dangler	Boomer Campbell	Bull Cook Ole

Calico Bill
Caribou Bill
Car Stake Smith
Charley the Logger
Clay Pipe
Clothes Pin Ole
Cordwood Johnson
Cougar Bill
Crosshaul Paddy
Cruel Face
Crying Charlie
Deacon
Dead Eye Dick
Dick the Dancer
Dirt Dan
Dirty Bill
Doc
Double Breasted
  Corrigan
Drop Cake Morley
Dutch Bill
Dynamite Bill
Eight Day Bill
Eight Day Tom
Farmer Dan
Frog Face
Gabby
Gin Pole Smith
Gypo Jack
Ham Bone Smith
Highpockets

Hobo Slim
Hungry Dan Shea
Jack the Horse
Jack the Ripper
Jessie James
Jimmie on the Trail
King Cole
Larry the Kicker
Lousy Dan
Mick
Minnesota Charley
Montana Slim
Moonlight Bob
Moonlight Pike
Moonshine Jimmie
Muskrat Moravetz
Needlenose Murphy
Old Dan
Old Hickory
Old Rampike
One Eye
One Round Hogan
Pancake Billie
Panicky Pete
Pastime Petey
Pembine Joe
Poker Peter
Protestant Jack
Prune Juice Doyle
Pump Handle
Rattlesnake Pete

Rollway Dick
Runaway Shea
Scotty
Silver Jack
Slippery Joe, the
  Mad Trapper
  from Rat River
Smutty John
Soo Line Kelley
Spike Maul
Spruce Hen Matt
Square Head
Squeaky George
Step and a-half
Stub Nelson
Stuttering Ed
The Bear
The Cleaver
The Hemlock Bull
The Pope
Three Day Max
Three Fingered Ole
Tin Can Murphy
Two Bit Shorty
Walking Daily
Whiskey Bill
Whispering Bill
Wicked Dan
Winnipeg Blackie
Wooden Shoes
Yellow Jacket Joe